LIFE-CYCLE OF STRUCTURES UNDER UNCERTAINTY

Emphasis on Fatigue-Sensitive Civil and Marine Structures

Dan M. Frangopol
Department of Civil and Environmental Engineering
ATLSS Engineering Research Center
Lehigh University, Bethlehem, Pennsylvania, USA

Sunyong Kim
Department of Civil and Environmental Engineering
Wonkwang University
Jeonbuk, Republic of Korea

CRC Press is an imprint of the
Taylor & Francis Group, an **informa** business
A SCIENCE PUBLISHERS BOOK

Cover credit: The cover photo is the San Francisco-Oakland Bay Bridge located in San Francisco, California, over the San Francisco Bay. The authors would like to thank Dr. Man-Chung Tang, T.Y. Lin International's Chairman of the Board, designer of more than 100 major bridges across the globe. The photo was taken by Engel Cheng/Shutterstock.com.

CRC Press
Taylor & Francis Group
6000 Broken Sound Parkway NW, Suite 300
Boca Raton, FL 33487-2742

First issued in paperback 2021

Version Date: 20190425

ISBN 13: 978-0-367-77940-5 (pbk)
ISBN 13: 978-0-367-14755-6 (hbk)

Library of Congress Cataloging-in-Publication Data

Names: Frangopol, Dan M., author. | Kim, Sunyong, 1976- author.
Title: Life-cycle of structures under uncertainty : emphasis on fatigue-sensitive civil and marine structures / Dan M. Frangopol (Department of Civil and Environmental Engineering, ATLSS Engineering Research Center, Lehigh University, Bethlehem, Pennsylvania, USA), Sunyong Kim (Department of Civil and Environmental Engineering, Wonkwang University, Jeonbuk, Republic of Korea).
Description: Boca Raton, FL : CRC Press, 2019. | "A science publishers book." | Includes bibliographical references and index.
Identifiers: LCCN 2019017167 | ISBN 9780367147556 (hardback)
Subjects: LCSH: Mechanical wear. | Materials--Fatigue. | Metals--Fatigue.
Classification: LCC TA418.4 .F73 2019 | DDC 620.1/126--dc23
LC record available at https://lccn.loc.gov/2019017167

Visit the Taylor & Francis Web site at
http://www.taylorandfrancis.com

and the CRC Press Web site at
http://www.crcpress.com

Preface

The performance, functionality and safety of civil and marine structures have been deteriorating due to age, increase in external loadings, corrosion, fatigue and other physical and chemical mechanisms. According to the various reports of the American Society of Civil Engineers published during the past three decades, the amount of deterioration in civil infrastructure has increased, the overall grade for the civil infrastructure in the United States has been degraded from C (fair) in 1988 to D (poor) in 2017, and the budget required to eliminate all existing and future structural deficiencies has grown substantially. As the age of a deteriorating structure increases and its service life reaches its design threshold life, there is a mounting risk associated with unsatisfactory performance under both normal and extreme loading conditions. The unexpected loss of functionality or failure of civil and marine structures can lead to severe economic, social and environmental impacts. For this reason, natural and financial resources should be allocated consistently and rationally to maintain the structural performance of deteriorating civil and marine structures above certain threshold levels.

Life-cycle analysis is a systematic tool for efficient and effective service life management of deteriorating structures. In the last few decades, theoretical and practical approaches for life-cycle performance and cost analysis have been developed extensively due to increased demand on structural safety and service life extension. This book presents the state-of-the-art in life-cycle analysis and maintenance optimization for fatigue-sensitive structures. Both theoretical background and practical applications have been provided for academics, engineers and researchers.

The primary topics covered in this book include (a) probabilistic concepts of life-cycle performance and cost analysis, (b) inspection and monitoring in life-cycle analysis, (c) fatigue crack detection under uncertainty, (d) optimum inspection and monitoring planning, (e) multi-objective life-cycle optimization, and (f) decision making in life-cycle analysis. For illustrative purposes, these topics have been applied to fatigue-sensitive details of bridges and ships.

This book includes eight chapters.

Chapter 1 provides the fundamental concepts of life-cycle analysis under uncertainty. Structural performance deterioration mechanisms such as corrosion and fatigue, the effects of maintenance actions on structural performance, cost and service life, and structural performance indicators related to structural condition, safety, tolerance to damage and cost are described. Furthermore, recent investigations on life-cycle optimization for fatigue-sensitive structures are reviewed.

Chapter 2 presents the role of inspection and monitoring in life-cycle analysis. This chapter covers the representative inspection, monitoring and maintenance methods for fatigue-sensitive structures, the effects of inspection and monitoring on life-cycle performance and cost under uncertainty, and statistical and probabilistic concepts associated with the efficient use of inspection and monitoring data including availability of monitoring data, loss function, Bayesian updating, and probabilistic importance indicators.

Chapter 3 describes the probabilistic concepts and methods related to fatigue crack damage detection, where the time-dependent fatigue crack propagation, probability of damage detection under multiple inspections, expected damage detection delay, and damage detection time-based probability of failure are provided. The concepts and approaches presented in this chapter are used for probabilistic optimum service life management in Chapters 4, 5 and 6.

In Chapter 4, the optimum inspection and monitoring planning for fatigue-sensitive structures is addressed using the fatigue crack damage detection-based objectives. The associated objectives are to maximize the lifetime probability of fatigue crack damage detection, minimize the expected fatigue crack damage detection delay, and minimize the fatigue crack damage detection time-based probability of failure. The formulations of these objectives are based on the probabilistic concepts provided in Chapter 3.

Chapter 5 deals with the optimum inspection and monitoring planning, considering the effects of inspection, monitoring and maintenance on service life extension and life-cycle cost. The objectives used in this chapter include minimizing the expected maintenance delay, maximizing the expected extended service life, and minimizing the expected life-cycle cost. The relationships among the number of inspections and monitorings, expected maintenance delay, expected extended service life and expected life-cycle cost are investigated.

Chapter 6 presents the multi-objective probabilistic optimum inspection and monitoring planning for fatigue-sensitive structures. The bi-, tri- and quad-objective optimization problems are investigated using the objective functions formulated in Chapters 4 and 5. Furthermore, the six objectives for optimum inspection planning and five objectives for optimum monitoring planning are used simultaneously to investigate the multi-objective optimization.

Chapter 7 addresses the decision making framework for optimum inspection and monitoring planning in order to deal with a large number of objectives efficiently, and to select the best single optimum inspection and monitoring plan for practical applications. In this framework, there are two decision alternatives such as decision making before and after solving multi-objective life-cycle optimization.

Chapter 8 serves as conclusions of this book. A summary of the book and future directions of the field of life-cycle performance and cost analysis and optimization for civil and marine structures under fatigue are also provided.

This book will help engineers engaged in civil and marine structures including students, researchers and practitioners with reliable and cost-effective maintenance planning of fatigue-sensitive structures, and to develop more advanced approaches and techniques in the field of life-cycle maintenance optimization and safety of structures under various aging and deteriorating conditions. Since the book is self-contained it can be used by all concerned with civil and marine structures, and probability and optimization concepts for fatigue-sensitive structures, including students, researchers and practitioners from all areas of engineering and industry. It can also be used for an advanced undergraduate course or a graduate course on life-cycle performance and cost of structures under uncertainty with emphasis on fatigue-sensitive structures. The areas to which the concepts and approaches presented in this book can be applied include not only civil structures, such as buildings, bridges, roads, railways, dams, and ports, and marine structures, such as naval vessels, offshore structures, submarines, submersibles, pipelines, and subsea systems, but also aerospace structures, nuclear power plants, and automotive structures.

The cover photo is the San Francisco-Oakland Bay Bridge located in San Francisco, California, over the San Francisco Bay. All the bridge components are constructed of steel. The authors would like to thank Dr. Man-Chung Tang, T.Y. Lin International's Chairman of the Board, designer of more than 100 major bridges across the globe. The photo was taken by Engel Cheng/Shutterstock.com.

Dan M. Frangopol
Department of Civil and Environmental Engineering
ATLSS Engineering Research Center, Lehigh University
117 ATLSS Drive, Bethlehem, PA 18015-4729, USA

Sunyong Kim
Department of Civil and Environmental Engineering
Wonkwang University, 460 Iksandae-ro, Iksan
Jeonbuk, 54538, Republic of Korea

Contents

Chapter **1**

Concepts of Life-Cycle Analysis Under Uncertainty

CONTENTS

ABSTRACT

Structural performance should be maintained above a certain threshold during the service life of a deteriorating structure. Financial resources for inspection, monitoring and maintenance have to be allocated consistently and rationally. Life-cycle performance and cost analysis can provide a practical solution to addressing the issues associated with the management of deteriorating structures under uncertainty. Chapter 1 presents the probabilistic concepts of life-cycle performance and cost analysis. This analysis requires an understanding of (a) structural performance under various deterioration mechanisms, (b) effects of maintenance actions on structural performance, cost and service life under uncertainty, and (c) structural performance indicators. The representative structural performance indicators associated with condition, safety, tolerance to damage and cost are described. Furthermore, the life-cycle optimization for service life management is also discussed. The objectives of life-cycle optimization are formulated based on the life-cycle performance and cost analysis. Finally, recent investigations into life-cycle optimization for fatigue-sensitive structures are addressed.

1.1 Introduction

Structural performance deteriorates over time due to aging, an increase in external loadings, corrosion, fatigue, and other physical and chemical mechanisms. As the age of a deteriorating structure increases and its service life reaches its design life, the risks of unserviceability and failure under normal and extreme loading conditions increase (Decò and Frangopol 2011, 2013; Ellingwood 2006). For this reason, structural performance should be maintained over certain thresholds during the service life of a deteriorating structure. Investments for inspection and maintenance have to be allocated consistently and rationally (ASCE 2017). The life-cycle performance and cost analysis can provide a rational and practical solution to address issues associated with the management of deteriorating structures (Estes and Frangopol 1997; Frangopol 2011, 2018; Frangopol et al. 2000; Frangopol and Soliman 2016; Frangopol et al. 2017; Kong and Frangopol 2003a, 2003b, 2005).

This chapter presents the probabilistic concepts of life-cycle performance and cost analysis. These concepts require an understanding of the (a) structural performance under various deterioration mechanisms (e.g., corrosion and fatigue), (b) effects of maintenance actions on structural performance, cost and service life, and (c) structural performance indicators under uncertainty. The structural performance indicators based on condition, safety, tolerance to damage and cost are reviewed. Furthermore, the life-cycle optimization for service life management is presented. The objectives of the life-cycle optimization are based on the life-cycle performance and cost analysis. Recent investigations on life-cycle optimization for fatigue-sensitive structures are addressed.

1.2 Life-Cycle Performance and Cost Analysis Under Uncertainty

Probabilistic life-cycle analysis leads to a systematic and rational evaluation of structural performance and cost. It serves as a basis for establishing an optimum service life management (Kim et al. 2013). Figure 1.1 shows the schematic for life-cycle analysis under uncertainty. The prediction of structural performance and cost under uncertainty is used to formulate the single- or multi-objective optimization process. Through this process, effective and efficient inspection and maintenance management can be achieved (Frangopol et al. 2012).

1.2.1 *Structural Performance Deterioration*

Structural performance prediction is one of the most significant phases in life-cycle analysis. Accurate prediction of the structural performance can result in the reliable service life management of deteriorating structures (Frangopol 2011, 2018). Structural performance deteriorates over time due

Figure 1.1 Schematic for life-cycle analysis under uncertainty.

Color version at the end of the book

to the combined effects of mechanical stressors, harsh environment, and extreme events. Gradual deterioration of the structural performance can be induced by corrosion and/or fatigue. These are the most common causes of resistance reduction in reinforced concrete (RC) and steel structures. Seismic events, hurricanes, floods and/or other extreme events may lead to a sudden drop in structural performance (Frangopol and Soliman 2016). During the past three decades, significant research towards predicting structural performance accurately has focused on development of time-dependent models of corrosion and fatigue initiation and propagation.

Corrosion has been considered one of the main factors causing deterioration of RC structures (NCHRP 2005, 2006). The corrosion process generally consists of corrosion initiation and propagation. Corrosion initiation can be defined as the time when the chloride concentration in the reinforcement

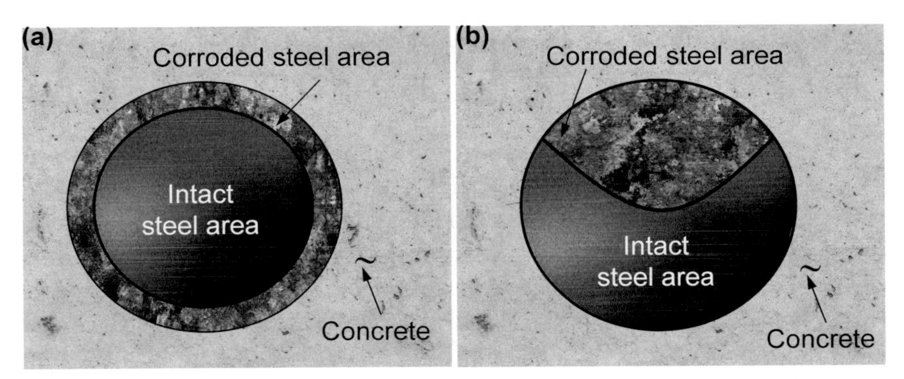

Figure 1.2 Cross sectional area reduction of steel in concrete due to corrosion: (a) uniform corrosion model; (b) pitting corrosion model.

steel surface reaches a predefined threshold (Arora et al. 1997; Stewart 2004; Zhang and Lounis 2006). Corrosion propagation produces damage such as cracking, spalling and reduction of reinforcement steel. The reduction of the reinforcement area can be represented by uniform and pitting corrosion models (Marsh and Frangopol 2008; Frangopol and Kim 2014a). As shown in Figure 1.2(a), the uniform corrosion model is based on the assumption that the cross sectional area of reinforcement steel is reduced uniformly. The pitting corrosion considers the highly localized corroded steel area as shown in Figure 1.2(b). Gonzalez et al. (1995), Stewart (2004) and Torres-Acosta and Martinez-Madrid (2003) conclude that pitting corrosion is more general for RC bridge decks, and results in a higher probability of failure and reduced service life than uniform corrosion when the same cross sectional area is considered.

Fatigue is one of the most common deterioration mechanisms in steel structures (Kwon and Frangopol 2011; Kwon et al. 2013). Under repeated loading and unloading, fatigue crack damage may initiate and accumulate at regions with initial flaws in the material, welding process or fabrication. When the number of cycles and stress ranges exceed their prescribed thresholds, the fatigue cracks can result in the fracture of the steel component and structural system failure. Details on fatigue damage propagation are addressed in Section 3.2.

1.2.2 Structural Performance and Cost with Maintenance

The structural performance of deteriorating structures can be maintained and improved by maintenance and replacement. As shown in Figure 1.3(a), the service life t_{life}, which is defined as the time when the structural performance reaches a predefined threshold P_{th}, can be extended by applying maintenance and replacement. The preventive maintenance (PM) is applied to maintain the performance above the required level (Kong and Frangopol 2003a, 2003b, 2005; Neves et al. 2006; Bocchini and Frangopol 2011; Frangopol and Kim 2014a). The PM application time t_{pm} can be predetermined considering the relation between improvement of the structural performance after the PM and cost required for the PM

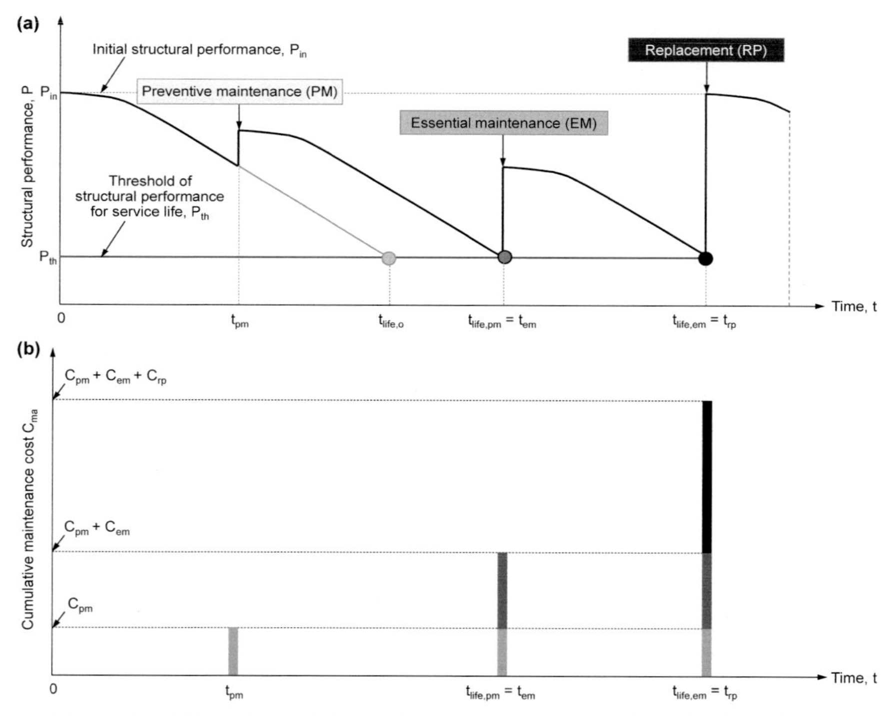

Figure 1.3 Life-cycle analysis considering maintenance actions: (a) deterministic structural performance and service life; (b) cumulative maintenance cost.

(Okasha and Frangopol 2010a). Service life can be extended from the initial service life $t_{life,0}$ to $t_{life,pm}$ by applying PM. At time t_{em}, when the structural performance reaches the threshold P_{th}, the essential maintenance (EM) is applied so that the service life can be extended from $t_{life,pm}$ to $t_{life,em}$. The structural performance returns to its initial state P_{in} by replacement (RP). Figure 1.3(b) shows the cumulative maintenance cost when the PM, EM and RP are applied at times t_{pm}, t_{em}, and t_{rp}, respectively. The PM can result in a relatively small improvement of structural performance. PM is associated with a much lower maintenance cost C_{pm} than those associated with EM and RP. It is relevant to note that PM can also result in no improvement of *P* but can delay the deteriorating process (e.g., painting of steel components). The RP leads to higher improvement of the structural performance *P* and larger cost C_{rp} than those associated with PM and EM.

Based on Figure 1.3, structural performance over time and extended service life under uncertainty are illustrated in Figure 1.4. The uncertainties associated with the improvement of *P* after PM, EM and RP can be represented by the associated probability density functions (PDFs). Furthermore, the PDFs of the extended service life by applying maintenance can be obtained using probabilistic concepts and methods. As mentioned previously, the probabilistic life-cycle analysis shown in Figure 1.4 is used for optimum service life management.

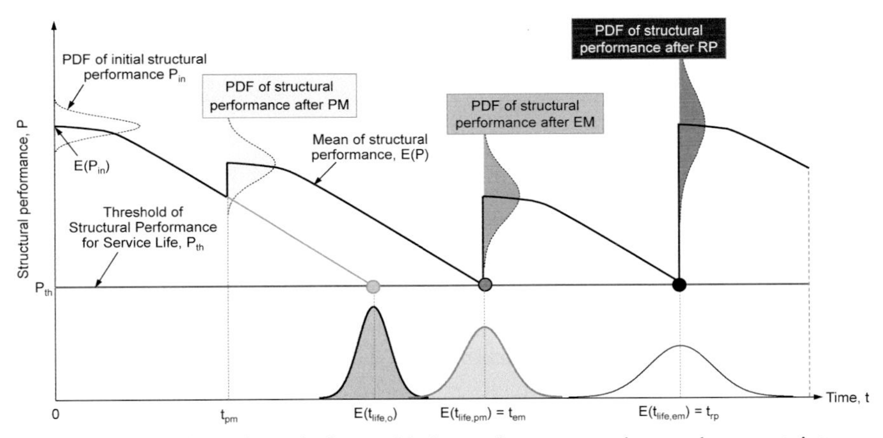

Figure 1.4 Life-cycle analysis considering maintenance actions under uncertainty.

1.3 Structural Performance Indicators

The structural condition, safety, tolerance to damage and cost can be represented by structural indicators (Zhu and Frangopol 2012; Frangopol and Saydam 2014; Ghosn et al. 2016a, 2016b; Biondini and Frangopol 2016). Condition rating is generally estimated based on visual inspection. In the United States, two representative condition rating methods are used for bridges. The first condition rating method is the National Bridge Inventory (NBI) condition rating, where the conditions of bridge components (e.g., deck, superstructures, substructures) are estimated using a value ranging from 0 to 9. The condition rating value of 0 is associated with the failed condition, and condition rating value of 9 indicates excellent condition. The second condition rating method is Pontis (Cambridge Systematics, Inc 2009), where the rating value ranges from 1 (i.e., no evidence of damage in a bridge component) to 5 (i.e., severe damage affecting the strength and serviceability of a bridge component). Table 1.1 describes the condition states of NBI and Pontis. More details on NBI and Pontis condition rating methods can be found in FHWA (1995), CDOT (1998), Estes and Frangopol (2003) and Saydam et al. (2013).

The representative safety-based structural performance indicators are the probability of failure P_f and reliability index β, which are expressed as

$$P_f = 1 - P_s = \int_{g(\mathbf{X})<0} f_{\mathbf{X}}(\mathbf{x})d\mathbf{x} \tag{1.1a}$$

$$\beta = \Phi^{-1}\left(1 - P_f\right) \tag{1.1b}$$

where $\mathbf{X}$ is a vector of random variables, $f_{\mathbf{X}}(\mathbf{x})$ is the joint PDF of random variables $\mathbf{X}$, and $\Phi^{-1}(\cdot)$ is the inverse of the standard normal cumulative distribution function (CDF). The state of the structure can be determined as follows: [$g(\mathbf{X}) > 0$] = "safe state"; [$g(\mathbf{X}) < 0$] = "failure state"; and [$g(\mathbf{X}) = 0$] = "limit state". If the random variables $\mathbf{X}$ of the state function $g(\mathbf{X})$ consist of the uncorrelated normal variables X_1 and X_2, the state of the structure can be illustrated as shown in Figure 1.5. The random variable X'_i in the standard normal space of Figure 1.5 is defined as $(X_i - \mu_i)/\sigma_i$, where μ_i and σ_i are the mean and standard deviation of X_i, respectively. In the standard

Table 1.1 NBI and Pontis condition ratings.

NBI condition rating for bridge deck, superstructure or substructure (adopted from FHWA (1995))		
Rating	**Conditions**	**Description**
N/A	Not applicable	Not applicable
9	Excellent	No problem noted
8	Very good	No problem noted
7	Good	Some minor problems
6	Satisfactory	Structural elements show some minor deterioration
5	Fair	All primary structural elements are sound but may have some minor section loss, cracking, spalling or scour
4	Poor	Advanced section loss, deterioration, spalling or scour
3	Serious	Loss of section, deterioration, spalling or scour have seriously affected primary structural components. Local failures are possible. Fatigue cracks in steel or shear cracks in concrete may be present
2	Critical	Advanced deterioration of primary structural elements. Fatigue cracks in steel or shear cracks in concrete may be present or scour may have removed substructure support
1	Imminent failure	Major deterioration or section loss present in critical structural components or obvious vertical or horizontal movement affecting structure stability
0	Failed	Out of service—beyond corrective action
Pontis condition rating for painted steel girder element (adopted from CDOT (1998))		
Rating	**Conditions**	**Description**
1	Good	There is no evidence of active corrosion and the paint system is sound and functioning as intended to protect the metal surface
2	Fair	There is little or no active corrosion. Surface or freckled rust has formed or is forming. The paint system may be chalking, peeling, curling or showing other early evidence of paint system distress but there is no exposure of metal

Table 1.1 contd. ...

...Table 1.1 contd.

Pontis condition rating for painted steel girder element (adopted from CDOT (1998))		
Rating	**Conditions**	**Description**
3	Paint failure	Surface or freckled rust is prevalent. There may be exposed metal but there is no active corrosion which is causing loss of section
4	Paint failure with steel corrosion	Corrosion may be present but any section loss due to active corrosion does not yet warrant structural analysis of either the element or the bridge
5	Major section loss	Corrosion has caused section loss and may be sufficient to warrant structural analysis to ascertain the impact on the ultimate strength and/or serviceability of either the element or the bridge

normal space, the reliability index β is the minimum distance from the origin to the limit state. The system reliability can be assessed through the appropriate system modeling. The probabilities of failure P_f of a series system, parallel system and series-parallel system in Figure 1.6 are expressed as

$$P_f = P\left(\bigcup_{i=1}^{N}\{g_i(\mathbf{X}) \le 0\}\right) \quad \text{for a series system} \qquad (1.2a)$$

$$P_f = P\left(\bigcap_{i=1}^{N}\{g_i(\mathbf{X}) \le 0\}\right) \quad \text{for a parallel system} \qquad (1.2b)$$

$$P_f = P\left(\bigcup_{i=1}^{M}\bigcap_{j=1}^{N_i}\{g_{ij}(\mathbf{X}) \le 0\}\right) \quad \text{for a series-parallel system} \qquad (1.2c)$$

where N = number of components in a series system (see Figure 1.6(a)), or a parallel system (see Figure 1.6(b)); M = number of parallel systems in a series-parallel system (see Figure 1.6(c)). The ith parallel system consists of N_i components. The probabilities of failure P_f of a series system for a perfectly correlated case and statistically independent case are expressed as

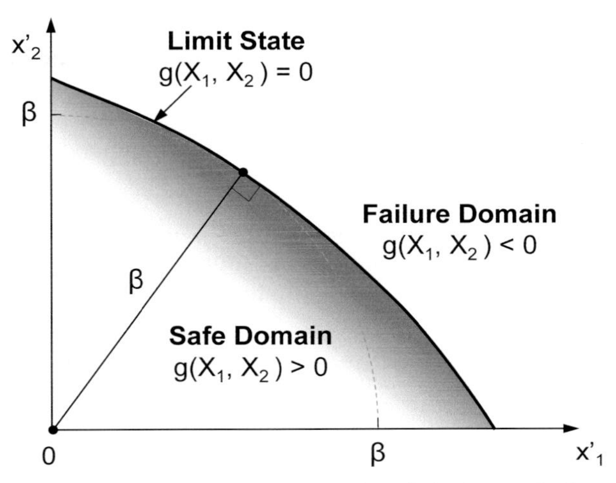

Figure 1.5 Safe domain, limit state and failure domain in the standard normal space.

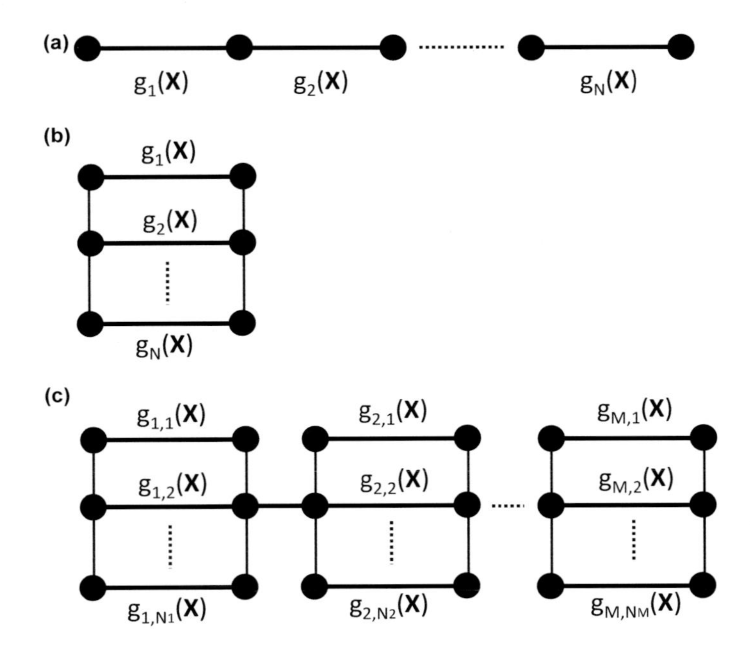

Figure 1.6 System modeling: (a) series system; (b) parallel system; (c) series-parallel system.

$$P_f = \max_{i=1}^{N}\{P_{f,i}\} \qquad \text{for perfectly correlated case} \quad (1.3a)$$

$$P_f = 1 - \prod_{i=1}^{N}(1 - P_{f,i}) \qquad \text{for statistically independent case} \quad (1.3b)$$

where $P_{f,i}$ is the probability of failure for the ith component in a series system. Furthermore, the probabilities of failure P_f of a parallel system for a perfectly correlated case and statistically independent case are computed as

$$P_f = \min_{i=1}^{N}\{P_{f,i}\} \qquad \text{for perfectly correlated case} \quad (1.4a)$$

$$P_f = \prod_{i=1}^{N} P_{f,i} \qquad \text{for statistically independent case} \quad (1.4b)$$

Equations (1.3) and (1.4) are associated with the two extreme cases (i.e., perfectly correlated case and statistically independent case). Considering the correlation among the random variables and various types of PDFs for the random variables, the probability of failure of the system can be computed (Cornell 1967; Ditlevsen 1979; Ang and Tang 1984; Thoft-Christensen and Murotsu 1986).

Structural redundancy, a measure of reserve capacity, has been used as a system level structural performance indicator for bridges and naval ships, among others (Frangopol and Curley 1987; Fu and Frangopol 1990; Frangopol and Nakib 1991; Hendawi and Frangopol 1994; Okasha and Frangopol 2010c; Ghosn et al. 2010; Decò et al. 2011, 2012; Zhu and Frangopol 2013c, 2015). Their structural redundancy can be quantified in several ways. Frangopol and Nakib (1991) expresses the structural redundancy index RI as

$$RI_1 = \frac{P_{f,dmg} - P_{f,sys}}{P_{f,sys}} \quad (1.5)$$

where $P_{f,dmg}$ is the probability of damage occurrence to the systems, and $P_{f,sys}$ is the probability of system failure. The difference between $P_{f,sys}$ and $P_{f,dmg}$ in Eq. (1.5) indicates the availability of system warning. If the damage occurrence to the system results in structural system failure (i.e., $P_{f,dmg}$ =

$P_{f,sys}$ in Eq. (1.5)), a structural system has no redundancy (i.e., $RI_I = 0$). Furthermore, the structural redundancy index RI can be expressed in terms of the reliability index as (Frangopol and Curley 1987)

$$RI_{II} = \frac{\beta_{int}}{\beta_{int} - \beta_{dmg}} \tag{1.6}$$

where β_{int} = reliability index of the intact system; β_{dmg} = reliability of the damaged system. When there is almost no change in reliability index β due to damage (i.e., $\beta_{int} \cong \beta_{dmg}$ in Eq. (1.6)), a structural system will have a very large structural redundancy index (i.e., $RI_{II} \to \infty$).

Robustness defines the ability of a structural system to resist progressive collapse under localised damage and damage tolerance of a structural system (Frangopol and Saydam 2014). According to Maes et al. (2006), the robustness of a structural system is expressed as

$$RB = \min_{i} \frac{P_{f,0}}{P_{f,i}} \tag{1.7}$$

where $P_{f,0}$ is the probability of failure of the undamaged system, and $P_{f,i}$ is the probability of failure of a system with the damaged component i. Furthermore, vulnerability can be used as the structural performance indicator to represent damage tolerance of a structure. Lind (1995) defines the vulnerability of a system VL as

$$VL = \frac{P_f\left(S_d, Q\right)}{P_f\left(S_0, Q\right)} \tag{1.8}$$

where P_f = probability of failure of the system; S_d = damage state d; S_0 = pristine system state; Q = prospective loading. Vulnerability $VL = 1$ indicates that the probability of failure of a pristine system is equal to the probability of failure of a damaged system.

Life-cycle cost is one of the most common cost-based structural performance indicators (Frangopol and Soliman 2016). The expected total life-cycle cost C_{life} can be computed as (Frangopol et al. 1997)

$$C_{life} = C_{int} + C_{insp} + C_{ma} + C_{fail} \tag{1.9}$$

where C_{int} is the initial cost; C_{insp} is the expected cost for inspections; C_{ma} is the expected cost for maintenance and repair actions; and C_{fail} is the expected failure cost. Risk, which is defined as multiplication of occurrence probability by the consequences (i.e., cost) of an event, can also be used to represent a structural performance. The formulation of the total risk is (Baker et al. 2008)

$$RS = P_{f,com,i} \cdot C_{dir,i} + P_{f,com,i} \cdot P_{f,subsys|com,i} \cdot C_{ind,i} \tag{1.10}$$

where $P_{f,com,i}$ = probability of failure of component i in a system; $P_{f,subsys|com,i}$ = probability of subsequent system failure to the failure of component i. $C_{dir,i}$ and $C_{ind,i}$ are the direct and indirect costs caused by the failure of component i. A system for which the indirect risk does not significantly affect the total risk is a robust system. Accordingly, the risk-based robustness index RB_{rs} is expressed as (Baker et al. 2008)

$$RB_{rs} = \frac{P_{f,com,i} \cdot C_{dir,i}}{P_{f,com,i} \cdot C_{dir,i} + P_{f,com,i} \cdot P_{f,subsys|com,i} \cdot C_{ind,i}} \tag{1.11}$$

The risk-based robustness index RB_{rs} ranges from 0 to 1. A larger value of RB_{rs} represents a higher robustness. The relation among the structural performance indicators defined in Eqs. (1.1) to (1.11), and the effects of deterioration mechanisms, loading and extreme events on the structural performance indicators were investigated in Okasha and Frangopol (2010a, 2010b, 2010c), Saydam and Frangopol (2011, 2013), Barone and Frangopol (2014), Zhu and Frangopol (2012, 2013a), Decò and Frangopol (2011, 2013), Decò et al. (2012), Dong and Frangopol (2016), Ghosn et al. (2016a, 2016b), among others.

1.4 Life-Cycle Optimization

Life-cycle optimization is performed for efficient service life management as shown in Figure 1.1. Through the life-cycle optimization process considering various objectives based on the structural performance, cost, and service life, the optimum times and types of inspection and maintenance applications are determined (Liu and Frangopol 2005b, 2006; Frangopol and Liu 2007; Frangopol and Kim 2014a). The relationship between expected structural performance during service life and present value of total expected life-cycle cost near the optimal region is illustrated in Figure 1.7. A larger expected structural performance can require a larger initial cost C_{int}, less expected failure cost C_{fail}, and less expected cost for inspection and maintenance $C_{insp} + C_{ma}$ during the service life of a structure. By applying the optimum inspection and maintenance strategy for the expected structural performance during the service life P^*, the expected total life-cycle cost C_{life} defined in Eq. (1.9) can be minimized to be C^*_{life} (Frangopol et al. 1997).

Optimization can be formulated with single and multiple objectives. It has been shown that as the number of objectives for the life-cycle optimization increases, more rational and well-balanced solutions can be obtained, and a more flexible decision can be made to select the best solution among the computed multiple solutions (Kim and Frangopol 2017, 2018a, 2018b). Figure 1.8 shows the feasible criterion space and the Pareto optimal set

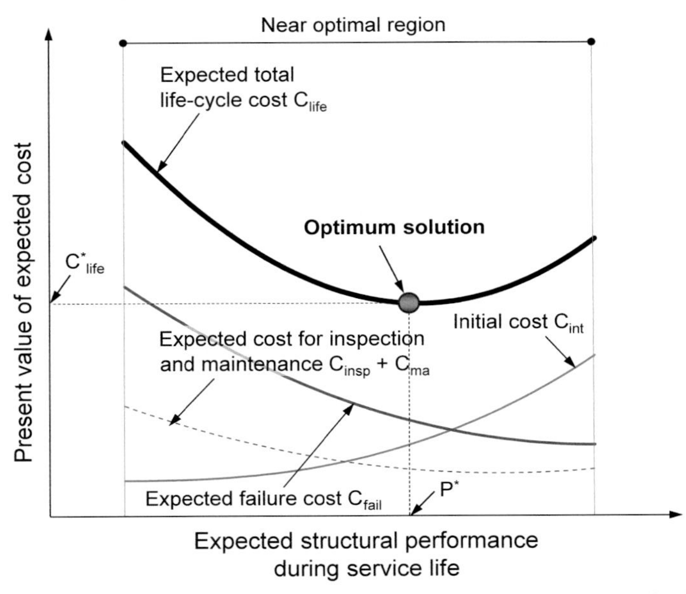

Figure 1.7 Relation between expected structural performance during service life and life-cycle cost near optimal region (Adapted from Frangopol (2011)).

when two objective functions f_1 and f_2 are considered simultaneously. The Pareto optimal set of the bi-objective optimization for minimizing both f_1 and f_2 is shown in Figure 1.8(a). When the bi-objective optimization is based on maximization of both f_1 and f_2, the Pareto optimal set is illustrated in Figure 1.8(b). Figure 1.8(c) shows the Pareto optimal set associated with minimization of f_1 and maximization of f_2. Finally, Figure 1.8(d) shows the Pareto optimal set associated with maximization of f_1 and minimization of f_2.

Recent investigations on life-cycle managements for deteriorating structures have adopted the multi-objective optimization. Table 1.2 summarizes recent investigations for fatigue-sensitive structures. The objectives used in the optimization are based on cost (e.g., expected life-cycle cost, maintenance cost, failure cost, inspection and monitoring cost),

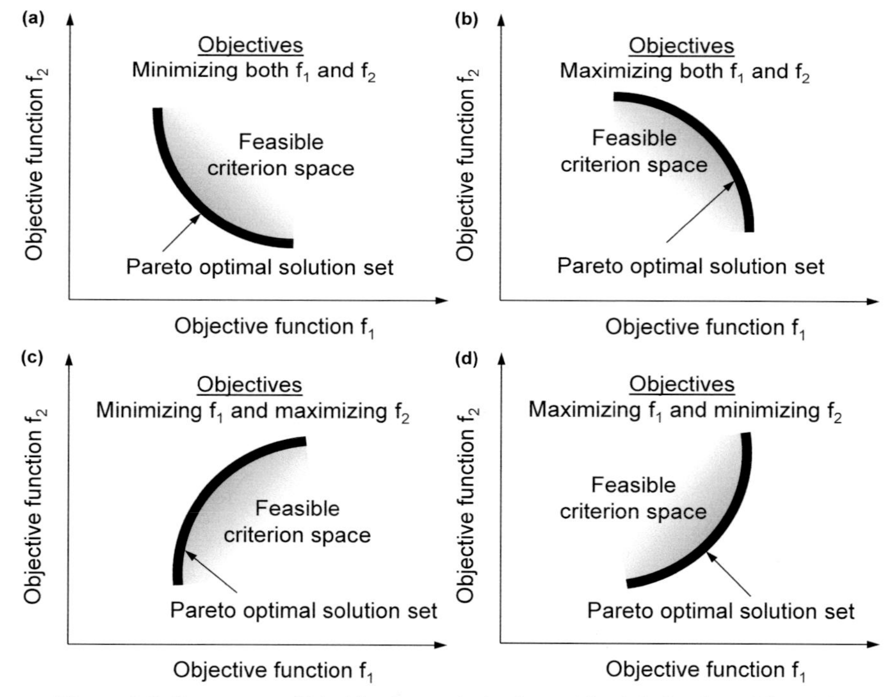

Figure 1.8 Pareto sets of bi-objective optimizations: (a) minimization of f_1 and f_2; (b) maximization of f_1 and f_2; (c) minimization of f_1 and maximization of f_2; (d) maximization of f_1 and minimization of f_2.

damage detection (e.g., probability of damage detection, expected damage detection delay, and maintenance delay), reliability index (or probability of failure), and service life. Through the life-cycle optimization, optimum inspection times and types (or quality), monitoring starting times and duration, and maintenance times and types (or quality) can be computed. Additional life-cycle optimizations for fatigue-sensitive structures are addressed in Chapters 4, 5 and 6.

Table 1.2 Recent investigations on life-cycle optimization for fatigue-sensitive structures.

Applications	Objectives	Design variables	References
Bridges	▪ Minimize expected total life-cycle cost	Inspection times and quality, and repair quality	Lukić and Cremona (2001)
	▪ Maximize reliability index ▪ Minimize retrofit size	Retrofit length and height	Liu et al. (2010a)
	▪ Minimize cumulative maintenance cost ▪ Minimize expected failure cost	Maintenance application times and types	Orcesi et al. (2010)
	▪ Maximize expected average availability of monitoring data ▪ Minimize total monitoring cost	Monitoring starting times and durations	Kim and Frangopol (2011b)
	▪ Minimize monitoring cost ▪ Minimize expected failure cost	Maintenance time and types	Orcesi and Frangopol (2011a)
	▪ Maximize probability of fatigue detection ▪ Minimize inspection cost	Inspection times and types	Soliman et al. (2013a)
	▪ Maximize probability of fatigue detection ▪ Minimize expected damage detection delay ▪ Minimize expected repair delay ▪ Minimize probability of failure ▪ Maximize expected extended service life ▪ Minimize expected total life-cycle cost	Inspection times	Kim and Frangopol (2018b)

Table 1.2 contd. ...

...Table 1.2 contd.

Applications	Objectives	Design variables	References
Ships	▪ Minimize expected total life-cycle cost	Inspection times and monitoring starting times	Kim and Frangopol (2011a)
	▪ Minimize expected damage detection delay ▪ Minimize inspection cost	Inspection times and quality	Kim and Frangopol (2011d)
	▪ Minimize expected damage detection delay ▪ Minimize inspection and monitoring cost	Inspection times and quality, monitoring times and durations, and combination of inspection and monitoring	Kim and Frangopol (2012)
	▪ Maximize expected extended service life ▪ Minimize expected total life-cycle cost	Inspection times and crack size for maintenance	Kim et al. (2013)
	▪ Maximize expected extended service life ▪ Minimize expected total life-cycle cost	Inspection times and types	Soliman et al. (2016)
	▪ Minimize expected damage detection delay ▪ Minimize expected maintenance delay ▪ Maximize reliability index ▪ Maximize expected service life extension ▪ Minimize expected total life-cycle cost	Monitoring starting times and durations	Kim and Frangopol (2018a)

1.5 Conclusions

In this chapter, general concepts of life-cycle performance and cost analysis under uncertainty were presented. The life-cycle structural performance and cost estimation should consider inspection and maintenance. The objectives used for the optimum service life management can be formulated based on the life-cycle structural performance and cost prediction. Through the single- or multi-objective optimization process, effective and efficient inspection and maintenance management can be achieved. The practical application of life-cycle performance and cost analysis requires accurate prediction of (a) structural performance using information obtained from inspection and structural health monitoring (Catbas et al. 2008, 2013; Decò and Frangopol 2015; Frangopol 2011; Frangopol et al. 2012; Okasha et al. 2010, 2011; Soliman et al. 2015; Frangopol and Soliman 2016), (b) service life extension by maintenance actions (Kim et al. 2011, 2013), and (c) structural performance loss due to extreme events such as earthquakes, tsunamis, floods and hurricanes (Akiyama et al. 2011, 2013; Dong and Frangopol 2017; Mondoro et al. 2017; Mondoro and Frangopol 2018).

Chapter **2**

Inspection and Monitoring in Life-Cycle Analysis

CONTENTS

ABSTRACT

Chapter 2 describes several representative techniques that are employed in inspection, monitoring, and maintenance processes for fatigue cracks. The effects of inspection and monitoring on life-cycle performance and cost under uncertainty are discussed. The performance assessment and prediction using inspection and monitoring data are described. The statistical and probabilistic concepts for the formulation of the availability of monitoring data are provided. The loss function is presented to quantify the monetary loss caused by the use of unavailable monitoring data and the performance prediction model. Furthermore, Bayesian updating based on information available from the inspections and monitoring is described. In this manner, the accuracy and effectiveness of life-cycle management for civil and marine structures can be improved. Probabilistic importance indicators of individual components in a structural system are investigated for efficient inspection and monitoring. The application of concepts and approaches presented lead to an accurate probabilistic life-cycle performance and cost analysis by integrating inspection and monitoring data efficiently.

2.1 Introduction

Life-cycle performance and cost analysis can lead to optimum service life management of civil and marine structures by integrating inspection, monitoring and maintenance in a rational way (Frangopol 2011, 2018;

Okasha and Frangopol 2011). Since the uncertainties associated with life-cycle analysis are unavoidable, efforts to treat these uncertainties effectively have been made with development of the approaches for life-cycle structural performance assessment and prediction (Frangopol 2011; Biondini and Frangopol 2016). Within the two types of uncertainty (i.e., aleatory and epistemic), the epistemic uncertainty can be reduced by improving the accuracy of the information and modeling required for life-cycle analysis (Frangopol et al. 2008a, 2008b; Strauss et al. 2008). Therefore, in recent years, studies related to the integration of inspection and monitoring data into life-cycle analysis have been conducted extensively (Catbas et al. 2008, 2013; Decò and Frangopol 2015; Frangopol et al. 2012; Frangopol 2018; Frangopol and Soliman 2016; Zhu and Frangopol 2013c).

In this chapter, the representative inspection, monitoring and maintenance methods for fatigue-sensitive structures are reviewed. The effects of inspection and monitoring on the life-cycle performance and cost under uncertainty are discussed. Performance assessment and prediction using inspection and monitoring data are described. The statistical and probabilistic concepts for the formulation of the availability of monitoring data are also provided. In addition, the formulation of the availability of monitoring data is based on the error between the monitoring data and the performance prediction model. In order to quantify the monetary loss caused by the use of unavailable monitoring data and the performance prediction model, the loss function is presented. Furthermore, Bayesian updating based on information from inspection and monitoring is provided. Probabilistic importance indicators of individual components in a structural system are defined for efficient inspection and monitoring. The concepts and approaches presented in this chapter can be used to integrate inspection and monitoring data into life-cycle analysis of civil and marine structures under uncertainty.

2.2 Inspection, Monitoring and Maintenance for Fatigue

Fatigue inspection plays a significant role in ensuring the safety and reliability of fatigue-sensitive steel structures (Soliman et al. 2016). The most commonly used inspection techniques for fatigue crack detection and identification include visual, magnetic, penetrant, ultrasonic and eddy current methods (Fisher et al. 1998; Moan 2005; Ciang et al. 2008; Staszewski 2008). A visual inspection is performed using human eyes with optical devices (e.g., telescopes, magnifying glass and fiberscope). This inspection is the simplest. The quality of visual inspection may be poor to detecting small cracks (Fisher et al. 1998). Magnetic particle inspection can detect discontinuity of a magnetic field in a steel plate by spraying fine magnetic particles. If discontinuities of the magnetic field due to cracks exist, resistance in magnetic field can increase. This method is effective for smooth surfaces. Accuracy of this method may be reduced when used on weld surfaces (Demsetz et al. 1996). Penetrant method uses a low viscosity and high capillary fluid to penetrate into surface cracks (Fisher et al. 1998). A liquid penetrant with red dye is applied to the surface after cleaning and a developer is sprayed to make cracks visible. This method is reliable for detection of surface cracks on a smooth surface. However, it is difficult to detect the crack propagation from weld toes. Ultrasonic inspection uses high frequency sound waves. If the ultrasonic wave encounters cracks, distortion in the reflected waves can be found. This method can be applied to steel plates with a thickness over 3 mm, and requires simple and portable devices and equipment (Fisher et al. 1998). However, this inspection method should be performed by well-trained inspectors having considerable experience (Hellier 2012; Soliman et al. 2016). Eddy current inspection is useful in detecting cracks near the surface in steel. Cracks in steel lead to changes in eddy currents produced through an electromagnetic induction. The eddy current inspection is able to detect cracks and to measure the thickness of painting quickly. The equipment for eddy current inspection is portable. However, it has a relatively high cost of equipment and well-experienced inspectors are needed (Demsetz et al. 1996; Hellier 2012).

An inspection to detect fatigue crack is generally limited to a single critical location and inappropriate for scanning all the fatigue critical locations efficiently. Monitoring systems can allow automated and long-term damage detection for multiple critical locations efficiently and reliably (Antonaci et al. 2012). The monitoring techniques for fatigue crack detection generally use acoustic emissions, ultrasonic guided waves, and Lamb waves (Ciang et al. 2008). There have been enormous developments associated with monitoring systems, monitoring data analysis, identification of fatigue cracks, and prognosis of fatigue crack propagation (Cho and Lissenden 2012). The selection of the most appropriate monitoring method should consider potential defects (e.g., corrosion and crack), materials and geometry of monitored structures, cost for installation, operation and maintenance of monitoring systems, and accuracy of monitored data, among others (Frangopol and Kim 2014a, 2014b, 2014c; Soliman et al. 2016).

Maintenance for fatigue cracks can be categorized as: (a) surface treatments; (b) repair of through-thickness cracks; and (c) modification of details or structures with cracks (FHWA 2013). The representative surface treatments are grinding, gas tungsten arc (GTA) of the weld toe, and impact treatments. The portion of a detail (e.g., edges of flanges or plates) containing small cracks can be removed by grinding. GTA remelting process can lead to removing the micro-discontinuities and reducing the stress concentration at the weld toe (Fisher et al. 1998). Impact treatment results in slow crack initiation and propagation by introducing compressive residual stress near the weld toe, reshaping the weld toe, and improving geometry. Hole drilling to arrest the through thickness fatigue cracks is one of the most widely used repair methods. A hole with sufficient diameter is placed at the tip of a crack to prevent crack growth. Drilled holes can be effectively used as a long-term maintenance as well as temporary maintenance if the holes are installed with an appropriate size and location (Connor and Lloyd 2017). Cut out and re-fabricated parts of components with thickness fatigue cracks are also effective repair methods in order to reproduce the same conditions as those before cracking (Fisher et al. 1998). Another method to repair the through thickness fatigue cracks is placing cover plates over the crack. It can be expected that the stress range will be reduced by adding the cross-sectional area of the cover plate (FHWA 2013). Details on the techniques for reducing the stress range associated with cracks are available in Fisher et al. (1998), FHWA (2013) and Connor and Lloyd (2017).

2.3 Inspection and Monitoring in Life-Cycle Analysis

Life-cycle analysis under uncertainty should be based on the accurate prediction of structural performance for practical applications. The uncertainty associated with the structural performance prediction can be reduced by using inspection and monitoring data appropriately and efficiently (Frangopol and Kim 2014a, 2014b, 2014c; Messervey et al. 2011; Okasha et al. 2010, 2011). For this reason, studies have focused on the efficient use of inspection and monitoring data as well as damage detection and identification using advanced NDE technologies (Frangopol 2011; Frangopol and Soliman 2016).

2.3.1 *Effect of Inspection and Monitoring on Life-Cycle Performance*

Inspection and monitoring are generally performed for damage detection and identification. If damage is not detected and identified, the information obtained from inspection and monitoring can be used to improve the accuracy of life-cycle performance analysis under uncertainty (Farhey 2005; Brownjohn 2007). Figure 2.1 shows the effect of inspection and monitoring on life-cycle performance analysis, where three cases (i.e., case A, case B, and case C) are considered for illustrative purposes. Case A is associated with the initial structural performance prediction without using inspection and monitoring data. Cases B and C represent the structural performance prediction by using inspection and monitoring data. Figure 2.1 is based on the assumptions that (a) the mean of initial structural performance $E(P_{in})$ deteriorates after reaching the mean of damage initiation time $E(t_{oc})$, and (b) the rate of structural performance deterioration is constant over time and updated using the data from inspection and monitoring at $E(t_{oc})$. As shown in Figure 2.1(a), the mean of performance deterioration rate $E(r_{p,A})$ for case A is less than the mean of initial performance deterioration rate $E(r_{p,B})$ associated with case B. $E(r_{p,A})$ is larger than the mean of performance deterioration rate $E(r_{p,C})$ for case C. The standard deviation of case A is assumed to be the largest among the three cases, since the use of inspection and monitoring data can result

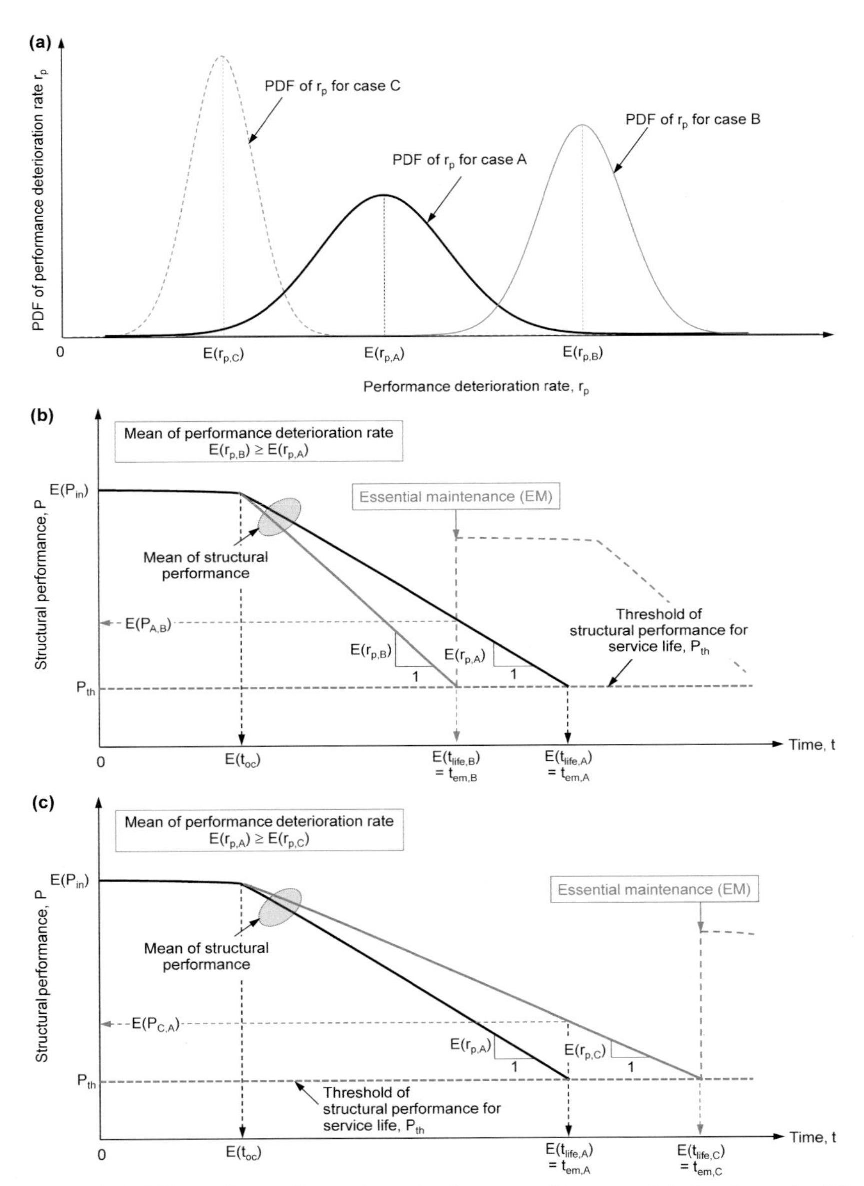

Figure 2.1 Effect of inspection and monitoring on performance deterioration rate: (a) PDFs of performance deterioration rate; (b) profiles of mean structural performance for cases A and B; (c) profiles of mean structural performance for cases A and C.

in reduction of the uncertainty associated with the structural performance deterioration rate r_p.

Figure 2.1(b) compares the profiles of the mean structural performance for cases A and B. The mean of the service life predicted with the updated structural performance deterioration rate $r_{p,B}$ will be smaller than the mean of the service life initially predicted (i.e., $E(t_{life,B}) < E(t_{life,A})$), since $E(r_{p,B})$ is larger than $E(r_{p,A})$ as indicated in Figure 2.1(a). In this case, the structural performance will reach a predefined threshold P_{th} earlier than initially predicted. Therefore, the essential maintenance, which is applied when the structural performance P reaches the threshold P_{th}, has to be performed earlier than the initially scheduled time (i.e., $t_{em,B} < t_{em,A}$). From Figure 2.1(b), it can be concluded that the appropriate updating based on inspection and monitoring information results in a timely maintenance action to prevent unacceptable structural performance, and finally reduces the probability of failure during the service life of deteriorating structures.

As shown in Figure 2.1(c), the mean of the updated structural performance deterioration rate $E(r_{p,C})$ is less than the mean of the initial performance deterioration rate $E(r_{p,A})$. The expected service life associated with $E(r_{p,C})$ may be larger than the expected initial service life with $E(r_{p,A})$ (i.e., $E(t_{life,C}) > E(t_{life,A})$). As a result, the essential maintenance can be applied later than the initially scheduled time (i.e., $t_{em,C} > t_{em,A}$). This implies that the appropriate updating process can lead to avoiding unnecessary maintenance action and reducing the maintenance cost. The performance deterioration rate of fatigue-sensitive civil and marine structures is related to parameters representing loading conditions, material properties, initial crack size, and geometry of the critical locations. The single or multiple parameters can be updated by Bayesian-based approaches with inspection and monitoring data (Zhu and Frangopol 2013b; Soliman and Frangopol 2014; Soliman et al. 2015; Liu and Frangopol 2019b).

2.3.2 *Effect of Inspection and Monitoring on Life-Cycle Cost*

The expected total life-cycle cost C_{life} generally consists of the initial cost C_{int}, cost for scheduled inspection C_{insp}, cost for maintenance and repair actions C_{ma}, and expected failure cost C_{fail}, as indicated in Eq. (1.9). By considering the cost for structural health monitoring (SHM) C_{mon}, the expected total life-cycle cost $C_{life,m}$ can be expressed as (Frangopol and Messervey 2011)

$$C_{life,m} = C_{int} + C_{insp} + C_{ma} + C_{fail} + C_{mon} \tag{2.1}$$

As shown in Figure 2.1, the updating based on information from SHM has an effect on the structural performance prediction, maintenance and service life. For this reason, C_{insp}, C_{ma} and C_{fail} required to formulate the expected total life-cycle cost may be affected by applying SHM. The cost-benefit of SHM C_{ben} can be estimated as

$$C_{ben} = C_{life} - C_{life,m} \tag{2.2}$$

where C_{life} = expected total life-cycle cost without SHM; $C_{life,m}$ = expected total life-cycle cost with SHM. Based on the relation between the expected structural performance and life-cycle cost presented in Figure 1.7, the expected total life-cycle costs with and without monitoring are compared in Figure 2.2. SHM requires the cost for installation, operation and maintenance of the monitoring system. If damage detection is more accurate by the effective SHM, it will imply less frequent in-depth inspection and maintenance. In this case, the cost for SHM can be compensated by the reduction in the cost of inspection and maintenance. Finally, the positive benefit of SHM C_{ben} can be obtained. On the contrary, if the effective SHM leads to more frequent in-depth inspection and maintenance, additional associated costs will be required. However, timely and appropriate maintenance based on more accurate (i.e., less uncertain) structural performance assessment and prediction will reduce the probability of failure and the expected failure cost C_{fail}. When the reduction of C_{fail} is larger than the increment in the cost for inspection, maintenance and SHM, the minimum expected total life-cycle cost $C^*_{life,m}$ associated with the effective SHM is less than the minimum expected total life-cycle cost C^*_{life} without SHM. Therefore, the benefit of SHM can be expected (i.e., $C_{ben} = C^*_{life} - C^*_{life,m} > 0$). This can be found from the comparison of Figures 2.2(a) and 2.2(b).

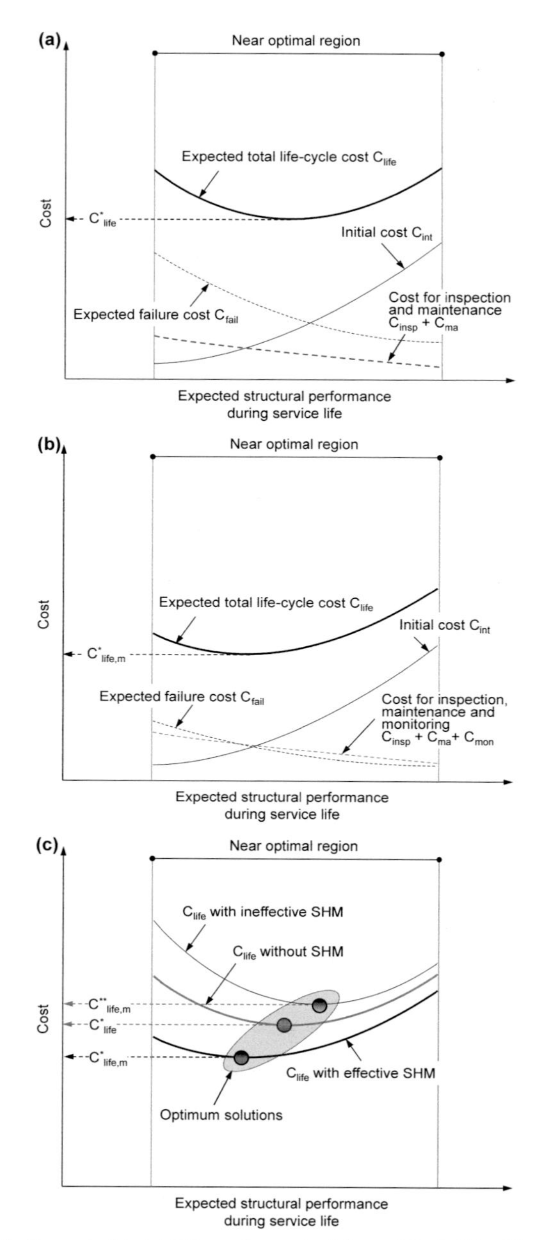

Figure 2.2 Expected total life-cycle costs: (a) no SHM; (b) effective SHM; (c) comparison among life-cycle costs.

If SHM is not effective, the increment in the cost for inspection, maintenance and SHM is larger than the reduction in the expected failure cost. Therefore, the minimum expected total life-cycle cost $C^{**}_{life,m}$ with ineffective SHM will be larger than C^{*}_{life} without SHM, and there will be a loss (i.e., $C_{ben} = C^{*}_{life} - C^{**}_{life,m} < 0$). Figure 2.2(c) compares the minimum expected total life-cycle costs C^{*}_{life}, $C^{*}_{life,m}$ and $C^{**}_{life,m}$. The effective SHM can be achieved through the successful integration of (a) optimum SHM system installation and management, (b) timely and accurate damage detection and identification, and (c) efficient use of SHM data for efficient and reliable life-cycle performance and cost analysis under uncertainty (Frangopol 2011; Frangopol and Kim 2014a; Frangopol and Soliman 2016).

2.4 Performance Prediction Using Inspection and Monitoring Data

Appropriate use of inspection and monitoring data can lead to (a) reduction of uncertainty and improvement of accuracy for the structural performance prediction, and (b) validation and updating of initial structural performance prediction model (Frangopol et al. 2008b; Strauss et al. 2008).

2.4.1 *Structural Performance Prediction Based on Inspection and Monitoring Data*

One of the most representative probabilistic structural performance indicators is the reliability as mentioned in Section 1.3. The time-dependent state function $g(t)$ for reliability analysis is generally expressed as

$$g(t) = S(t) - L(t) \tag{2.3}$$

where S is the structural capacity of a structure, and L is the load effect. S and L are treated as time-dependent random variables. Using the inspection and monitoring data, the random variables S and/or L in Eq. (2.3) can be estimated. Frangopol et al. (2008a) proposed the approach for the efficient inclusion of monitoring data in the structural reliability assessment, where the mean and standard deviation of the strain measured during the long term monitoring are used to estimate the probabilistic characteristics of random variable L in Eq. (2.3). Frangopol et al. (2008b) and Strauss et al. (2008) developed probabilistic approaches to predict the structural reliability of bridges based on monitoring extreme data (i.e., daily maximum stress) induced by traffic loads, where the regression and acceptance sampling approaches are applied to predict the daily maximum stress. The procedure for assessing and predicting the bridge component and system performance was proposed by Liu et al. (2009a, 2009b). In their studies, the structural performance is represented by the probability that the monitored stress exceeds the predefined stress threshold. Fatigue reliability assessment using monitoring data for steel bridges and naval ship structures can be found in Liu et al. (2010b), Kwon and Frangopol (2010), and Kwon et al. (2013). In their fatigue reliability assessment, the limit state function is formulated as

$$g(t) = N_c - N(t) \tag{2.4}$$

where N_c is the total number of cycles to fatigue failure, which is determined by the stress range-number of cycles (S-N) relationship; and $N(t)$ is the accumulated number of stress cycles to the time t, which is determined from monitoring stress ranges and associated number of cycles.

2.4.2 *Availability of Inspection and Monitoring Data*

When monitoring is performed continuously for a long period of time, and all the data from monitoring is recorded and saved, a large data storage system is required. In order to deal with the monitored data efficiently, the statistics of extremes can be used (Messervey et al. 2011; Frangopol and Kim 2014c). The extreme values from the monitored data may be treated as random. Assuming that the cumulative distribution functions (CDFs) of the n initial random variables ($X_1, X_2, \ldots, X_n$) are identical (i.e., $F_{X_1}(x) = F_{X_2}(x) = \ldots = F_{X_n}(x)$) and statistically independent, the CDF of the largest value among n samples $F_{Y_{max}}(y)$ is expressed as (Ang and Tang 2007)

$$F_{Y_{max}}(y) = P(X_1 \leq y, X_2 \leq y, \ldots, X_n \leq y) = [F_X(y)]^n \tag{2.5}$$

where $Y_{max} = \max(X_1, X_2, \ldots, X_n)$. As $n \to \infty$, $F_{Ymax}(y)$ will converge to an asymptotic distribution. According to Gumbel (1958), the asymptotic distribution can be categorized into (a) double exponential form; (b) single exponential form; and (c) exponential form with an upper (or lower) boundary.

The availability of monitoring data for structural performance prediction is defined as the probability that the performance prediction model can be used during a prediction period (Kim and Frangopol 2010, 2011b; Sabatino and Frangopol 2017). The formulation of the availability of monitoring data considers the residuals between monitoring data and the regression function as shown in Figure 2.3. Assuming that the residuals are normally distributed with zero mean, the daily maximum positive residual can be represented by the double exponential form of the asymptotic distribution. Based on the probability of exceedance, the probability that the largest positive residual in future non-monitoring t_{nm} days is larger than the largest positive residual during monitoring period t_m days $P_{ex}(t_{nm})$ is (Ang and Tang 2007)

$$P_{ex}\left(t_{nm}\right) = 1 - exp\left(-\frac{t_{nm}}{t_m}\right) \tag{2.6}$$

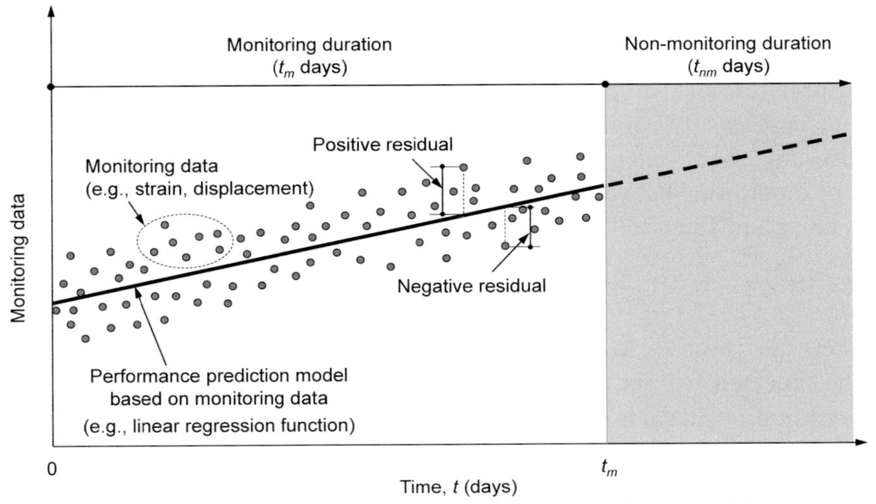

Figure 2.3 Monitored data and the performance prediction model.

The criterion employed for the availability of monitoring data is as follows: if the largest positive residual for the non-monitoring t_{nm} days exceeds the largest positive residual during the monitoring t_m days, the monitoring data cannot be used for structural performance prediction. Accordingly, the availability of monitoring data $A(t_{nm})$ during the non-monitoring t_{nm} days can be expressed as

$$A(t_{nm}) = 1 - P_{ex}(t_{nm}) \tag{2.7}$$

Considering the allowable number of future exceedances N_{al}, the availability of monitoring data can be formulated as indicated in Table 2.1 (Kim and Frangopol 2011b). Cases A1, A2 and A3 are associated with $N_{al} = 0$, 1 and 2, respectively, considering only the largest positive residual value. Equation (2.7) is based on case A1 with the allowable number of future exceedances $N_{al} = 0$. The availability of monitoring data for case A2 with $N_{al} = 1$ is associated with the criterion that if the largest positive residual for the non-monitoring t_{nm} days exceeds the largest positive residual during the monitoring t_m days at least two times, the monitoring data cannot be used for future performance prediction. Cases B1, B2 and B3 consider both the largest positive residual and the smallest negative residual. The associated allowable number of future exceedances N_{al} is one, two and three, respectively. When the largest positive residual (or the smallest negative residual) during the non-monitoring t_{nm} days exceeds the largest positive residual (or the smallest negative residual) during the monitoring t_m days at least two times, the monitoring data is not usable for prediction, according to the criterion for case B2 with $N_{al} = 1$. The average availability $\bar{A}$ of monitoring data during the non-monitoring t_{nm} days can be expressed as (Ang and Tang 2007)

$$\bar{A} = A(t_{nm}) + \frac{t_l}{t_{nm}} \cdot (1 - A(t_{nm})) \qquad \text{for } 0 \leq t_l < t_{nm} \tag{2.8}$$

where t_l = time to lose usability of the monitoring data for prediction. The expected average availability $E(\bar{A})$ of monitoring data during the non-monitoring t_{nm} days is

$$E(\bar{A}) = \frac{1}{t_{nm}} \int_o^{t_{nm}} A(t)\, dt \tag{2.9}$$

Table 2.1 Availability of monitoring data for structural performance prediction considering allowable number of exceedances.

Cases	Allowable number of exceedances N_{al} for available monitoring data	Residual to be considered	Availability $A(t_{nm})$
A1	$N_{al} = 0$	Largest positive residual	$exp\left(-\frac{t_{nm}}{t_m}\right)$
A2	$N_{al} = 1$	Largest positive residual	$\frac{(t_{nm}+t_m)}{t_m}\cdot exp\left(-\frac{t_{nm}}{t_m}\right)$
A3	$N_{al} = 2$	Largest positive residual	$\left\{1+\frac{t_{nm}}{t_m}+\frac{1}{2}\left(\frac{t_{nm}}{t_m}\right)^2\right\}\cdot exp\left(-\frac{t_{nm}}{t_m}\right)$
B1	$N_{al} = 0$	Both largest positive residual and smallest negative residual	$exp\left(-\frac{2t_{nm}}{t_m}\right)$
B2	$N_{al} = 1$	Both largest positive residual and smallest negative residual	$\left(\frac{t_{nm}+t_m}{t_m}\right)^2\cdot exp\left(-\frac{2t_{nm}}{t_m}\right)$
B3	$N_{al} = 2$	Both largest positive residual and smallest negative residual	$exp\left(-\frac{2t_{nm}}{t_m}\right)\cdot\left\{\frac{t_{nm}{}^4}{4t_m^4}+\frac{t_{nm}{}^3}{t_m^3}+\frac{2t_{nm}{}^2}{t_m^2}+\frac{2t_{nm}}{t_m}+1\right\}$

Figure 2.4 shows the relation between the expected average availability $E(\bar{A})$ and ratio of monitoring duration to non-monitoring duration (i.e., t_m/t_{nm}) for cases A1, A2, A3, B1, B2 and B3 in Table 2.1. As shown in

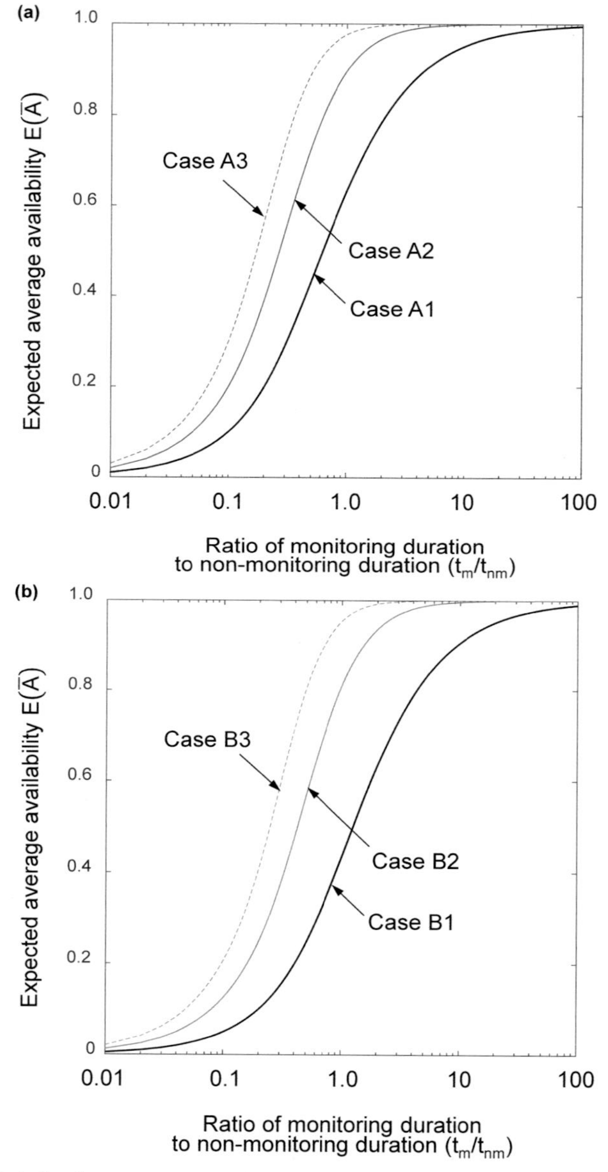

Figure 2.4 Relation between expected average availability of monitoring data to ratio of monitoring duration to non-monitoring duration: (a) cases A1, A2 and A3; (b) cases B1, B2 and B3.

Figure 2.4(a), the expected average availability $E(\bar{A})$ for case A1 is less than that associated with cases A2 or A3. This is because the availability of case A1 is less than that associated with cases A2 or A3 (see Table 2.1). Similarly, the expected average availabilities $E(\bar{A})$ for case B1 provides the least values as shown in Figure 2.4(b), since the availability of case B1 is less than that associated with cases B2 or B3. The details on theoretical background and applications of availability of monitoring data for future performance predictions can be found in Kim and Frangopol (2010, 2011b).

2.4.3 Structural Performance Prediction Error and Loss

The structural performance prediction model based on monitoring data may be unavailable because of errors. The unavailable performance prediction model can lead to monetary loss related to untimely and inappropriate maintenance actions and unexpected loss of serviceability of civil and marine structures (Frangopol and Kim 2014c). This monetary loss can be quantified based on the loss function (Taguchi et al. 1988). The loss function $f_{loss}(x, \hat{x})$ can be expressed as (Ang and Tang 1984)

$$f_{loss}(x, \hat{x}) = C_{loss}(x - \hat{x})^2 \tag{2.10}$$

where x = actual value of parameter; $\hat{x}$ = predicted value of parameter; and C_{loss} = monetary loss caused by an error in prediction. The expected loss $E_{loss}(x, \hat{x})$ is estimated as

$$E_{loss}\left(x,\hat{x}\right) = \int_{-\infty}^{\infty} f_{loss}\left(x,\hat{x}\right) \cdot f_X\left(x\right)\, dx = C_{loss}\left\{\sigma_x{}^2 + \left(\mu_x - \hat{x}\right)^2\right\} \tag{2.11}$$

where σ_x and μ_x are the standard deviation and mean of x, respectively. Equation (2.11) shows that $E_{loss}(x, \hat{x})$ is reduced by decreasing the standard deviation σ_x and the difference between μ_x and $\hat{x}$.

Considering the relation between the sample size and cost of sampling, the loss function of Eq. (2.10) can be modified as

$$f_{loss}(x, \hat{x}, n_{sp}) = C_{loss}(x - \hat{x})^2 + C_{sp} \cdot n_{sp} \tag{2.12}$$

where n_{sp} = number of samples; and C_{sp} = cost per sample. Depending on whether prior information exists or not, the optimum sample size n^*_{sp} required to minimize the expected loss E_{loss} is (Ang and Tang 1984)

$$n^*_{sp} = \sigma_x \cdot \sqrt{\frac{C_{loss}}{C_{sp}}} - \left(\frac{\sigma_x}{\sigma'_x}\right)^2 \geq 0 \qquad \textit{with prior information} \quad (2.13a)$$

$$n^*_{sp} = \sigma_x \cdot \sqrt{\frac{C_{loss}}{C_{sp}}} \qquad \textit{without prior information} \quad (2.13b)$$

where σ'_x = prior standard deviation of the parameter x. It should be noted that Eq. (2.13) is available when x is normally distributed. For example, suppose that the loss function associated with performance prediction is $f_{loss}(r, \hat{r}, n_{sp}) = C_{loss}(r - \hat{r})^2 + C_{sp} \cdot t_m$, where r is the residual between the value from the performance prediction model and the representative daily monitoring data (see Figure 2.3). When the standard deviation of residual $\sigma_r = 2$, the relation among optimum monitoring days t^*_m, prior standard deviation of residual σ'_r and ratio of monetary loss to cost per sample (i.e., C_{loss}/C_{sp}) is shown in Figure 2.5. It can be found that (a) a larger monetary loss C_{loss} relatively to C_{sp} requires a larger number of monitoring days t^*_m, (b) t^*_m with prior information is less than that without prior information for given C_{loss}/C_{sp}, and (c) prior information with a smaller prior standard deviation of residual σ'_r results in a smaller t^*_m.

2.4.4 *Updating Based on Inspection and Monitoring Data*

Updating based on information from inspection and monitoring is essential to reduce the uncertainty associated with the structural performance assessment and prediction and to improve the accuracy and effectiveness of life-cycle management for civil and marine structures (Enright and Frangopol 1999b; Frangopol and Soliman 2016). The Bayesian inference concepts have been applied to update the parameters of the models for structural performance prediction. When the random variable X is described by its probability of density function (PDF) $f_X(x)$, and the

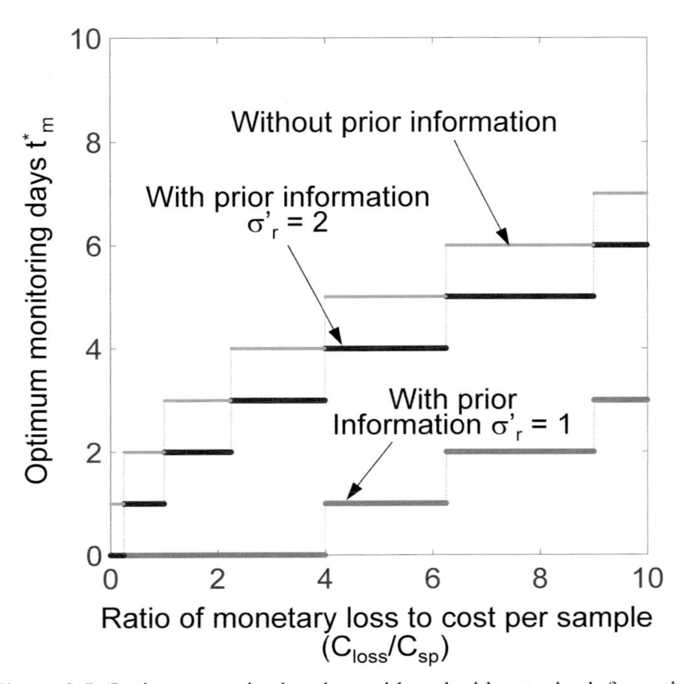

Figure 2.5 Optimum monitoring days with and without prior information.

parameter θ for the PDF $f_X(x)$ is treated as a random variable, the PDF of the initial parameter θ can be updated to the posterior PDF using the likelihood function $L(\theta)$. The posterior PDF $f''(\theta)$ of random variable θ can be found as (Ang and Tang 2007)

$$f''(\theta) = \frac{L(\theta) \cdot f'(\theta)}{\int_{-\infty}^{\infty} L(\theta) \cdot f'(\theta) d\theta} \tag{2.14}$$

where $f'(\theta)$ is the prior PDF of random variable θ. The new information consisting of N discrete sample values is used to construct $L(\theta)$ as

$$L(\theta) = \prod_{i=1}^{N} f_X(x_i | \theta) \tag{2.15}$$

where $f_X(x_i|\theta)$ is the PDF of random variable X estimated at the inspection and monitoring data value x_i for given parameter θ of the PDF $f_X(x)$. The posterior PDF $f''_X(x)$ of X is expressed as (Bucher 2009)

$$f''_X(x) = \int_{-\infty}^{\infty} f_X(x|\theta) \cdot f''(\theta) d\theta \tag{2.16}$$

For example, suppose that (a) the PDF $f_X(x)$ of X is normally distributed; (b) the initial probabilistic parameter θ is the mean of $f_X(x)$; (c) θ is normally distributed with the mean and standard deviation of 0 and 2, respectively (denoted as $N(0; 2)$); and (d) the additional information from inspection and monitoring provides the likelihood function $L(\theta)$ associated with $N(5; 3)$. According to Eq. (2.14), the posterior PDF $f''(\theta)$ of the probabilistic parameter θ can be obtained as shown in Figure 2.6(a). The updated PDF $f''(\theta)$ is characterized by $N(1.54; 1.66)$. If the standard deviation of the initial PDF of X is known as 3.0, the posterior PDF $f''_X(x)$ of X can be obtained using Eq. (2.16), as shown in Figure 2.6(b). From a comparison of $f''_X(x)$ and $f''(\theta)$ in Figure 2.6(b), it can be seen that the random variables X and θ have the same mean value, and the standard deviation of X is larger than that of θ. When the two parameters θ_1 and θ_2, which are used to characterize the PDF $f_X(x)$, are considered simultaneously for updating, the posterior PDF $f''(\theta_1, \theta_2)$ can be expressed as

$$f''(\theta_1, \theta_2) = \frac{L(\theta_1, \theta_2) \cdot f'(\theta_1, \theta_2)}{\int_{-\infty}^{\infty} \int_{-\infty}^{\infty} L(\theta_1, \theta_2) \cdot f'(\theta_1, \theta_2) d\theta_1 \cdot d\theta_2} \tag{2.17}$$

The likelihood function $L(\theta_1, \theta_2)$ is

$$L(\theta_1, \theta_2) = \prod_{i=1}^{N} f_X(x_i|\theta_1, \theta_2) \tag{2.18}$$

Therefore, the updated posterior PDF of X becomes (Zhu and Frangopol 2013b)

$$f''_X(x) = \int_{-\infty}^{\infty} \int_{-\infty}^{\infty} f_X(x|\theta_1, \theta_2) \cdot f''(\theta_1, \theta_2) d\theta_1 \cdot d\theta_2 \tag{2.19}$$

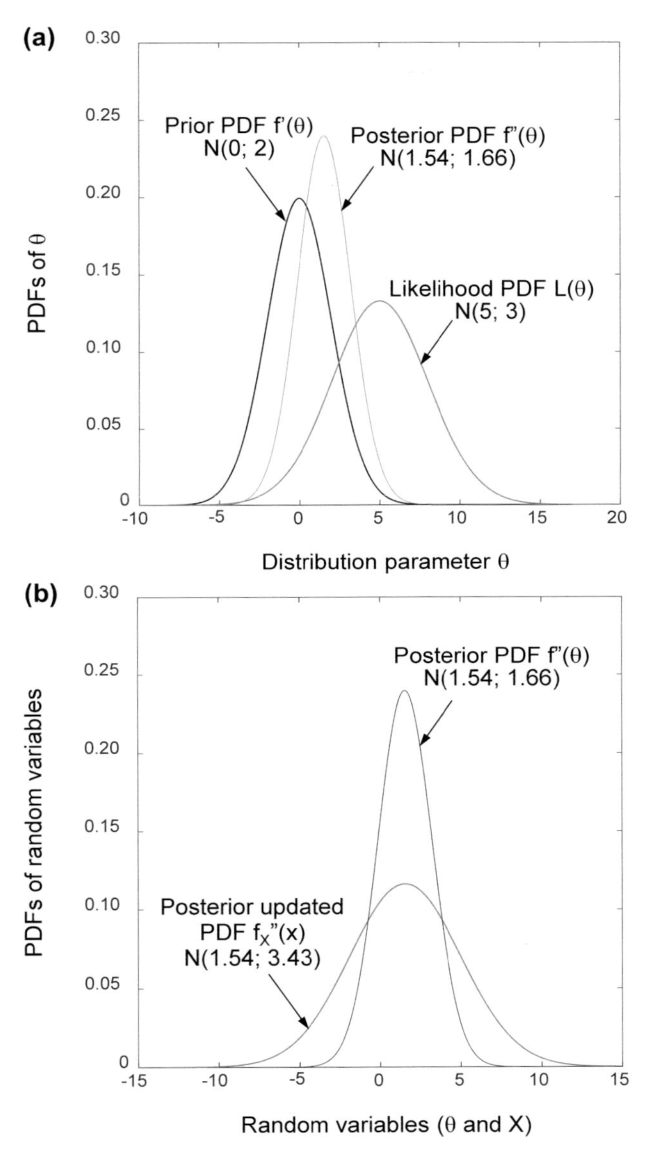

Figure 2.6 Bayesian updating: (a) prior, likelihood and posterior PDFs of probabilistic parameter θ; (b) posterior PDFs of random variables θ and X.

The Bayesian updating-based approaches for integration of inspection and monitoring data have been widely used for life-cycle performance assessments of civil and marine structures. Okasha and Frangopol (2012) proposed a general framework for bridge management using SHM data, where the Bayesian updating based on simulation methods (e.g., Metropolis-Hasting algorithm and slice sampling algorithm) were used for an extreme value distribution representing the traffic load effects. In Zhu and Frangopol (2013b), updating multiple parameters of the Rayleigh and extreme value distributions with the information acquired from monitoring was investigated to reduce the uncertainty in the performance assessment of ship structures. Soliman and Frangopol (2014) showed that Bayesian updating can be used to find the updated parameters of the fatigue crack propagation model, and the optimum inspection time can be determined using the updated parameters.

2.5 Importance Factors for Efficient Inspection and Monitoring

The structural system performance is estimated considering the performances of all the individual components in the system. In terms of the contribution of an individual component to the system performance, some components may be more critical than others (Leemis 2009; Modarres et al. 2017). The appropriate importance measures for individual components can lead to efficient inspection, maintenance and monitoring (Liu and Frangopol 2005a; Kim and Frangopol 2010). The Birnbaum importance factor is one of the most representative reliability-based importance factors, which is defined as (Birnbaum 1969)

$$I_{b,i} = \frac{\partial P_{s,sys}\left(P_s\right)}{\partial P_{s,i}} \tag{2.20}$$

where $I_{b,i}$ = Birnbaum importance factor of component i; $P_{s,sys}$ = reliability of the system as a function of the reliability of its individual components P_s; and $P_{s,i}$ = reliability of the ith component. Larger $I_{b,i}$ corresponds to more importance. If the individual components are independent, $I_{b,i}$ can be estimated as (Hoyland and Rausand 1994)

$$I_{b,i} = P_{s,sys}(P_s|P_{s,i} = 1) - P_{s,sys}(P_s|P_{s,i} = 0) \tag{2.21}$$

where $P_{s,sys}(P_s|P_{s,i} = 1)$ and $P_{s,sys}(P_s|P_{s,i} = 0)$ are the values of system reliability associated with $P_{s,i}$ equal to 1 and 0, respectively. Considering both the Birnbaum importance factor $I_{b,i}$ and the ratio of the reliability of an component $P_{s,i}$ to the system reliability $P_{s,sys}$, the criticality importance factor $I_{cr,i}$ of component i is defined as (Modarres et al. 2017)

$$I_{cr,i} = I_{b,i} \times \frac{P_{s,i}}{P_{s,sys}} \tag{2.22}$$

The contribution of component i to the system failure, which is called the Fussell-Vesely importance factor, can be expressed as (Fussell 1975)

$$I_{fv,i} = \frac{P_i}{P_{f,sys}} = \frac{1 - \prod_{j=1}^{m} P_{i,j}}{P_{f,sys}} \tag{2.23}$$

where P_i is the probability that the ith component is contributing to the system failure, m is the number of minimal cut sets that contain the ith component, and $P_{i,j}$ is the probability that the jth cut set containing the ith component is failed. Furthermore, the risk-reduction worth (RRW) importance factor $I_{rrw,i}$ and risk achievement worth (RAW) importance factor $I_{raw,i}$ of the ith component are (Ericson 2015; Modarres et al. 2017), respectively

$$I_{rrw,i} = P_{f,sys} - P_{f,sys}(P_f|P_{f,i} = 0) = P_{s,sys}(P_s|P_{s,i} = 1) - P_{s,sys} \tag{2.24}$$

$$I_{raw,i} = P_{f,sys}(P_f|P_{f,i} = 1) - P_{f,sys} = P_{s,sys} - P_{s,sys}(P_s|P_{s,i} = 0) \tag{2.25}$$

where $P_{f,sys}(P_f|P_{f,i} = 0)$ and $P_{f,sys}(P_f|P_{f,i} = 1)$ are the values of the system failure when $P_{f,i}$ is equal to 0 and 1, respectively. The normalized importance factor I_i^{norm} is

$$I_i^{norm} = \frac{I_i}{\sum_{i=1}^{n} I_i} \tag{2.26}$$

where I_i = importance factor of the ith component computed from Eqs. (2.20) to (2.25); and n = number of individual components in a system.

For example, suppose that a series-parallel system consists of three independent components (see Figure 2.7(a)), and the reliabilities of these components $P_{s,1}$, $P_{s,2}$ and $P_{s,3}$ are 0.98, 0.97 and 0.99, respectively. Using Eqs. (1.3) and (1.4), the system reliability $P_{s,sys}$ can be formulated as $\{1-(1-P_{s,1})(1-P_{s,2})\}P_{s,3}$. $P_{s,sys}(P_s|P_{s,1}=1)$ and $P_{s,sys}(P_s|P_{s,1}=0)$ are computed as 0.99 and 0.96, respectively. Based on Eq. (2.21), the Birnbaum importance factor $I_{b,1}$ of the component 1 becomes 0.99 – 0.96 = 0.03. In this manner, the Birnbaum importance factors $I_{b,2}$ and $I_{b,3}$ can be computed as 0.02 and 1.0, respectively. Therefore, the normalized Birnbaum importance factors I_b^{norm} for the three components are 0.028, 0.019 and 0.953. Table 2.2 provides the normalized importance factors of the three components based on the Birnbaum, criticality, Fussell-Vesely, RRW and RAW importance measures (i.e., I_b^{norm}, I_{cr}^{norm}, I_{fv}^{norm}, I_{rrw}^{norm} and I_{raw}^{norm}, respectively). When the financial resources for inspection and monitoring are limited, component 3 can take priority over components 1 and 2 for inspection and monitoring.

If the time-dependent reliabilities of the three components in Figure 2.7(a) are expressed as $P_{s,1}(t) = 0.98 - 0.03t$, $P_{s,2}(t) = 0.97 - 0.04t$, and $P_{s,3}(t) = 0.99 - 0.001t$, the normalized importance factors I_b^{norm}, I_{cr}^{norm}, I_{fv}^{norm}, I_{rrw}^{norm} and I_{raw}^{norm} over time are illustrated in Figures 2.7(b), 2.7(c), 2.7(d), 2.7(e) and 2.7(f), respectively. Figure 2.7(b) shows that component 3 has the largest normalized importance factor I_b^{norm} from 0 year to 16 years, and after 16 years the largest I_b^{norm} is associated with component 1. According to the criticality, Fussell-Vesely, and RAW importance measures, the normalized importance factor of component 3 is the largest, and the normalized importance factor of component 2 is the smallest over 20 years, as shown in Figures 2.7(c), 2.7(d), and 2.7(f). The normalized RRW importance factors of components 1 and 2 (i.e., $I_{rrw,1}^{norm}$ and $I_{rrw,2}^{norm}$) are identical, and are larger than $I_{rrw,3}^{norm}$ after 3 years (see Figure 2.7(e)).

Table 2.2 Normalized importance factors of individual components in Figure 2.7(a).

	Reliability of individual component $P_{s,i}$	**Birnbaum importance factor $I_{b,i}$**	**Criticality importance factor $I_{cr,i}$**	**Fussell-Vesely importance factor $I_{fv,i}$**	**RRW importance factor I_{rrwi}**	**RAW importance factor I_{rawi}**
Component 1	0.98	0.028	0.028	0.028	0.053	0.028
Component 2	0.97	0.019	0.019	0.019	0.053	0.019
Component 3	0.99	0.953	0.953	0.953	0.894	0.953

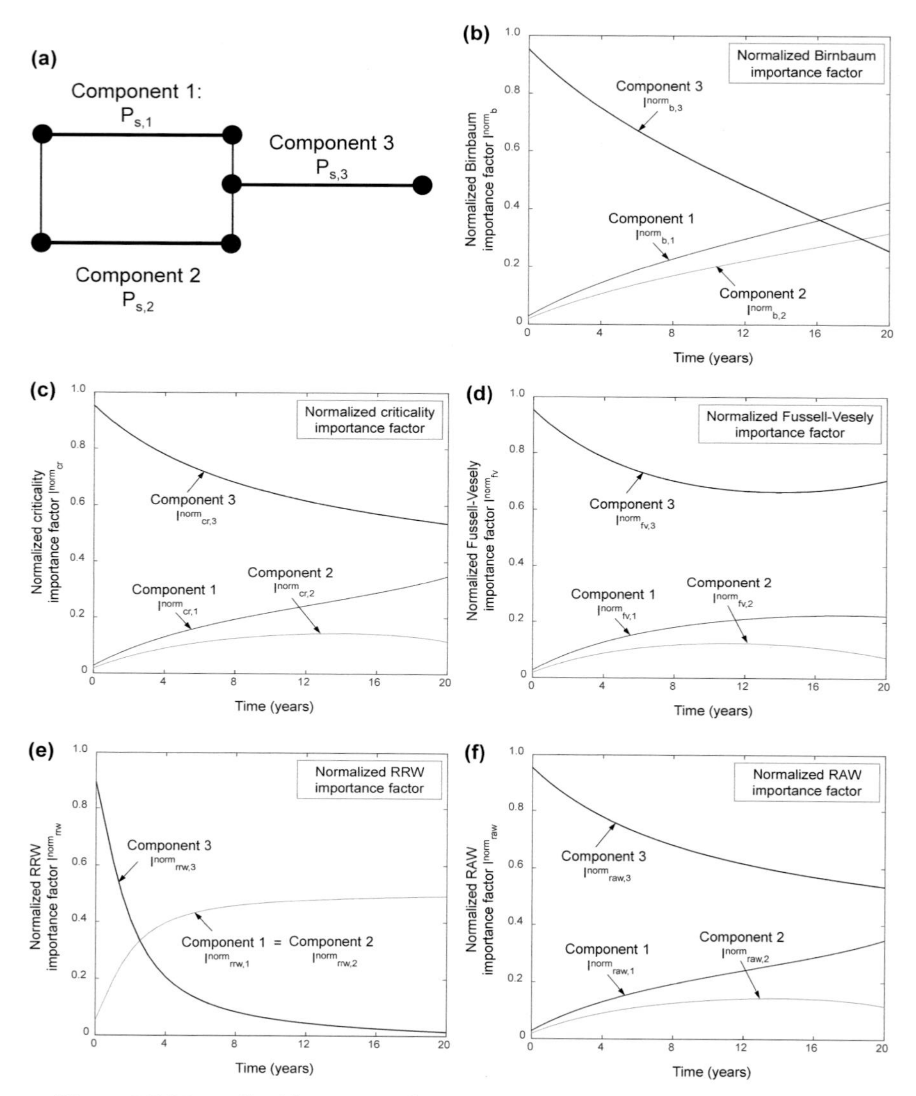

Figure 2.7 Normalized importance factor: (a) Series-parallel system; (b) Birnbaum importance; (c) Criticality importance; (d) Fussell-Vesely importance; (e) RRW importance; (f) RAW importance.

Based on the normalized importance factors for the three components in Figure 2.7, the annual inspection and monitoring cost for a structural system can be allocated. For example, if the total annual inspection and monitoring cost $C^{an}_{insp} + C^{an}_{mon}$ is \$20,000, and the inspection and monitoring cost of the ith component is assigned in proportion to the annual mean of the normalized Birnbaum importance factor $E[I^{norm}_{b,i}(t)]$, the inspection and monitoring cost of the ith component $C^{an}_{insp,i}(t) + C^{an}_{mon,i}(t)$ from $(t - 1)$ years to t years is computed as

$$C^{an}_{insp,i}(t) + C^{an}_{mon,i}(t) = (C^{an}_{insp} + C^{an}_{mon})\, E[I^{norm}_{b,i}(t)] \tag{2.27}$$

$E[I^{norm}_{b,i}(t)]$ is the mean of the normalized Birnbaum importance factor from $(t - 1)$ years to t years, and is computed from Figure 2.7(b). Table 2.3 provides $E[I^{norm}_{b,i}(t)]$ of components 1, 2 and 3 for $t = 1$ to 20. $E[I^{norm}_{b,i}(1)]$ of components 1, 2 and 3 for $t = 1$ (i.e., from 0 to 1 year) are 0.0455, 0.0319 and 0.9226, respectively. Therefore, the annual inspection and monitoring cost $C^{an}_{insp,1}(1) + C^{an}_{mon,1}(1)$ of component 1 becomes $0.0455 \times \$20{,}000 = \$910$ during the first year. $C^{an}_{insp,i}(1) + C^{an}_{mon,i}(1)$ for components 2 and 3 are computed as $0.0319 \times \$20{,}000 = \$637$ and $0.9226 \times \$20{,}000 = \$18{,}453$, respectively. In this manner, $E[I^{norm}_{b,i}(20)]$ of components 1, 2 and 3 for $t = 20$ can be also estimated as $0.4174 \times \$20{,}000 = \$8{,}348$, $0.3118 \times \$20{,}000 = \$6{,}236$ and $0.2708 \times \$20{,}000 = \$5{,}416$, respectively (see Table 2.3). Figure 2.8 shows the assigned annual inspection and monitoring cost $C^{an}_{insp,i}(t) + C^{an}_{mon,i}(t)$ of components 1, 2 and 3 for $t = 1$ year to 20 years. With the allocated inspection and monitoring cost, the times and types of inspection and/or monitoring can be optimized (Kim and Frangopol 2010).

Table 2.3 Annual mean of the normalized Birnbaum importance factors $E[I^{norm}_{b,i}(t)]$ of components 1, 2 and 3.

Years	t	$E[I^{norm}_{b,1}(t)]$	$E[I^{norm}_{b,2}(t)]$	$E[I^{norm}_{b,3}(t)]$
0–1	1	0.0455	0.0319	0.9226
1–2	2	0.0774	0.0559	0.8667
2–3	3	0.106	0.0775	0.8165
3–4	4	0.132	0.097	0.771
4–5	5	0.1559	0.115	0.7291
5–6	6	0.178	0.1317	0.6902
6–7	7	0.1988	0.1474	0.6538
7–8	8	0.2184	0.1622	0.6194
8–9	9	0.2371	0.1762	0.5866
9–10	10	0.2551	0.1897	0.5552
10–11	11	0.2724	0.2028	0.5249
11–12	12	0.2892	0.2154	0.4954
12–13	13	0.3057	0.2278	0.4665
13–14	14	0.3218	0.24	0.4382
14–15	15	0.3378	0.252	0.4102
15–16	16	0.3537	0.2639	0.3824
16–17	17	0.3695	0.2758	0.3546
17–18	18	0.3854	0.2877	0.3269
18–19	19	0.4013	0.2997	0.299
19–20	20	0.4174	0.3118	0.2708

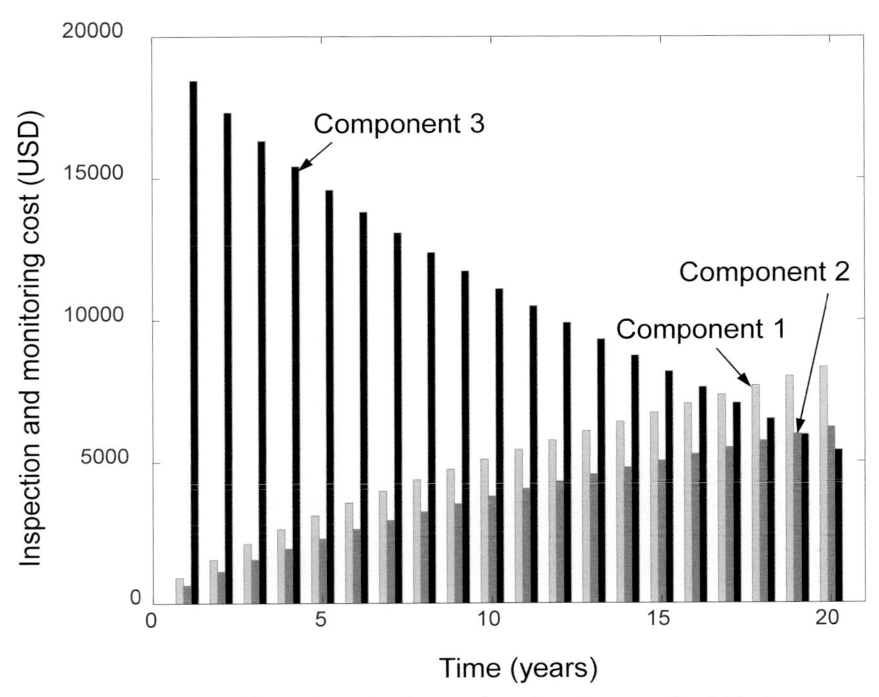

Figure 2.8 Inspection and monitoring cost based on the normalized Birnbaum importance factor.

2.6 Conclusions

This chapter describes several representative techniques for inspection, monitoring and maintenance for fatigue cracks. The effects of inspection and monitoring on life-cycle performance and cost analysis under uncertainty were investigated. For efficient use of inspection and monitoring data, the probabilistic concepts of structural performance assessment and prediction, and availability of inspection and monitoring data were presented. Based on these concepts, the optimum monitoring scheduling can be established through the multi-objective optimization by maximizing the expected average availability and minimizing the monitoring cost (Kim and Frangopol 2010, 2011b). In order to quantify the monetary loss induced by use of unavailable inspection and monitoring data, the loss function was introduced. Bayesian updating using information from inspection and monitoring reduces the uncertainty associated with structural performance assessment and prediction, and improves the accuracy and effectiveness of life-cycle management for civil and marine structures. The probabilistic importance factors can help managers determine efficient inspection and monitoring prioritization of structural components. The application of concepts and approaches presented in this chapter leads to accurate and reliable probabilistic life-cycle performance and cost analysis by integrating inspection and monitoring data efficiently.

Chapter 3

Probabilistic Damage Detection

CONTENTS

ABSTRACT

Chapter 3 presents the probabilistic concepts and methods related to fatigue crack damage detection. An overview of time-dependent crack propagation is provided to predict the crack size at a specific time and the time for the crack to reach a predefined size. The relation between the probability of fatigue crack damage detection and crack size is presented. The probability of damage detection for multiple inspections is formulated. Uncertainties associated with the damage occurrence time and damage detection are considered in the formulation of the expected damage detection delay. Furthermore, using the damage detection delay and a time-based safety margin, the damage detection time-based probability of failure is formulated. The effects of inspection times on the probability of fatigue damage detection, damage detection delay and damage detection time-based probability of failure are investigated. These concepts and approaches are presented in this chapter, including the formulation of damage detection-based optimization objectives, and can be used for probabilistic optimum inspection and monitoring planning in Chapter 4, 5, 6 and 7.

3.1 Introduction

Service life management of fatigue sensitive structures involves many uncertainties associated with the prediction of fatigue crack growth and damage detection (Garbatov and Guedes Soares 2001, 2014; Kim and Frangopol 2011a; Kwon and Frangopol 2010, 2012; Liu and Frangopol 2019a, 2019b; Soliman et al. 2013b). Appropriate and rational estimation of fatigue crack damage growth, based on probabilistic techniques can lead to efficient and effective life-cycle cost management and improvement of the structural safety of fatigue sensitive structures (Kim et al. 2013; Soliman et al. 2016). For this reason, the probabilistic concepts and methods related to fatigue damage detection have been considered as a significant and attractive tool for the life-cycle management of deteriorating structures under fatigue (Akpan et al. 2002; Frangopol 2011; Frangopol et al. 2012; Frangopol and Soliman 2016; Liu and Frangopol 2019a, 2019b). Therefore, understanding of probabilistic fatigue damage detection assessment is necessary and essential for structural engineers and decision makers.

This chapter presents probabilistic concepts and methods related to fatigue crack damage detection. An overview of time-dependent crack propagation is provided to predict the crack size at a specific time and time for the crack to reach a predefined crack size. The relation between the probability of fatigue crack damage detection and crack size is presented. The probability of damage detection for multiple inspections is also formulated. The damage occurrence time and damage detection under uncertainty are integrated in the formulation of the expected damage detection delay. Furthermore, the damage detection time-based probability of failure is formulated using the damage detection delay and time-based safety margin. The concepts and approaches presented in this chapter can be used for probabilistic optimum service life management in the remaining chapters.

3.2 Time-Dependent Fatigue Damage Propagation

Fatigue cracks may be pre-existing from the fabrication of steel members. The cracks can also be initiated and propagated under repetitive loads. The location and size of initial cracks of a steel member, stress range, number of cycles, material properties and geometry conditions of the member can affect crack propagation. Since these factors interact with each other in complex ways, it is difficult to predict crack size accurately. Among several empirical and phenomenological-based crack propagation models, Paris' law, which is based on linear elastic fracture mechanics, has been generally used (Paris and Erdogan 1963)

$$\frac{da}{dN_{cy}} = C\left(\Delta K\right)^m \qquad \text{for } \Delta K > \Delta K_{th} \tag{3.1}$$

where a = crack size; N_{cy} = cumulative number of cycles; ΔK = stress intensity factor; and ΔK_{th} = threshold of stress intensity factor. C and m are material crack propagation parameters. The stress intensity factor ΔK in Eq. (3.1) is (Irwin 1958)

$$\Delta K = S_r \cdot Y(a) \cdot \sqrt{\pi a} \tag{3.2}$$

where S_r = stress range; and $Y(a)$ = geometry function. From Eqs. (3.1) and (3.2), the cumulative number of cycles N_{cy} resulting in crack size a_N is estimated as (Fisher 1984)

$$N_{cy} = \frac{1}{C \cdot S_r^m} \int_{a_o}^{a_N} \left[Y(a) \sqrt{\pi a} \right]^{-m} da \tag{3.3}$$

where a_0 = initial crack size. Considering the annual number of cycles N_{an} and increase rate of number of cycles r_{cy}, the time t to attain the crack size a_n can be predicted as (Madsen et al. 1985, 1991; Kim and Frangopol 2012)

$$t = \frac{ln[1 + \frac{ln(1 + r_{cy})}{N_{an} \cdot C \cdot S_r{}^m} \cdot \int_{a_o}^{a_N} \left[Y(a)\sqrt{\pi \cdot a} \right]^{-m} da]}{ln(1 + r_{cy})} \quad \text{for } r_c > 0 \tag{3.4a}$$

$$t = \frac{1}{N_{an} \cdot C \cdot S_r{}^m} \cdot \int_{a_0}^{a_N} \left[\mathrm{Y}(a)\sqrt{\pi \cdot a} \right]^{-m} da \quad \text{for } r_c = 0 \tag{3.4b}$$

If the geometry function $Y(a)$ is constant (i.e., $Y(a) = Y$), and the annual number of cycles N_{an} is constant (i.e., $r_c = 0$ in Eq. (3.4), see Eq. (3.4b)), the time t for a crack to propagate from the initial crack length a_0 to the crack size a_N can be obtained as (Guedes Soares and Garbatov 1996a, 1996b; Kim and Frangopol 2011d)

$$t = \frac{a_N{}^{(2-m)/2} - a_o{}^{(2-m)/2}}{(\frac{2-m}{2}) \cdot C \cdot S_r{}^m \cdot Y^m \cdot \pi^{m/2} \cdot N_{an}} \quad \text{for } m \neq 2 \tag{3.5a}$$

$$t = \frac{ln(a_N) - ln(a_o)}{C \cdot S_r{}^m \cdot Y^m \cdot \pi \cdot N_{an}} \quad \text{for } m = 2 \tag{3.5b}$$

Furthermore, the crack length after N_{cy} cycles a_N is computed from Eqs. (3.1) and (3.2) as (Madsen et al. 1985; Kim and Frangopol 2011d)

$$a_N = [a_0^{(2-m)/2} + (\frac{2-m}{2}) \cdot C \cdot S_r^m \cdot Y^m \cdot \pi^{m/2} \cdot N_{cy}]^{(\frac{2}{2-m})} \quad \text{for } m \neq 2 \qquad (3.6a)$$

$$a_N = a_0 \cdot exp[C \cdot S_r^m \cdot Y^m \cdot \pi \cdot N_{cy}] \quad \text{for } m = 2 \qquad (3.6b)$$

The variables associated with Eqs. (3.4), (3.5) and (3.6) are randomized by using Monte Carlo simulation such that the probability density function (PDF) of time t to attain the specific crack size a_N can be obtained. As an illustrative example, the joint between the bottom plate and longitudinal plate of a ship hull subjected to fatigue is investigated in this chapter. The associated probabilistic descriptors of the variables in Eqs. (3.4), (3.5) and (3.6) are provided in Table 3.1. Detailed information on this example is available in Kim and Frangopol (2011d). Figure 3.1(a) shows the time-dependent crack size with the PDFs of time t for a_N = 10, 20, 30, and 40 mm. The PDF of time for the crack size to reach 1 *mm* is illustrated in Figure 3.1(b), where the mean and standard deviation of this time are 4.29 years and 2.63 years, respectively.

Table 3.1 Variables for crack growth model at the joint between bottom plate and longitudinal plate of a ship hull.

Variables	**Notation**	**Units**	**Mean**	***COV**	**Type of distribution**
Initial crack size	a_o	mm	0.5	0.2	Lognormal
Annual number of cycles	N_{an}	cycles/ year	0.8 × 10^6	0.2	Lognormal
Stress range	S_r	MPa	40	0.1	Weibull
Material crack growth parameter	C	-	3.54 × 10^{-11}	0.3	Lognormal
Material exponent	m	-	2.54	-	Deterministic
Geometry function value	Y	-	1.0	-	Deterministic

*COV: coefficient of variation.
Based on information in Kim and Frangopol (2011d).

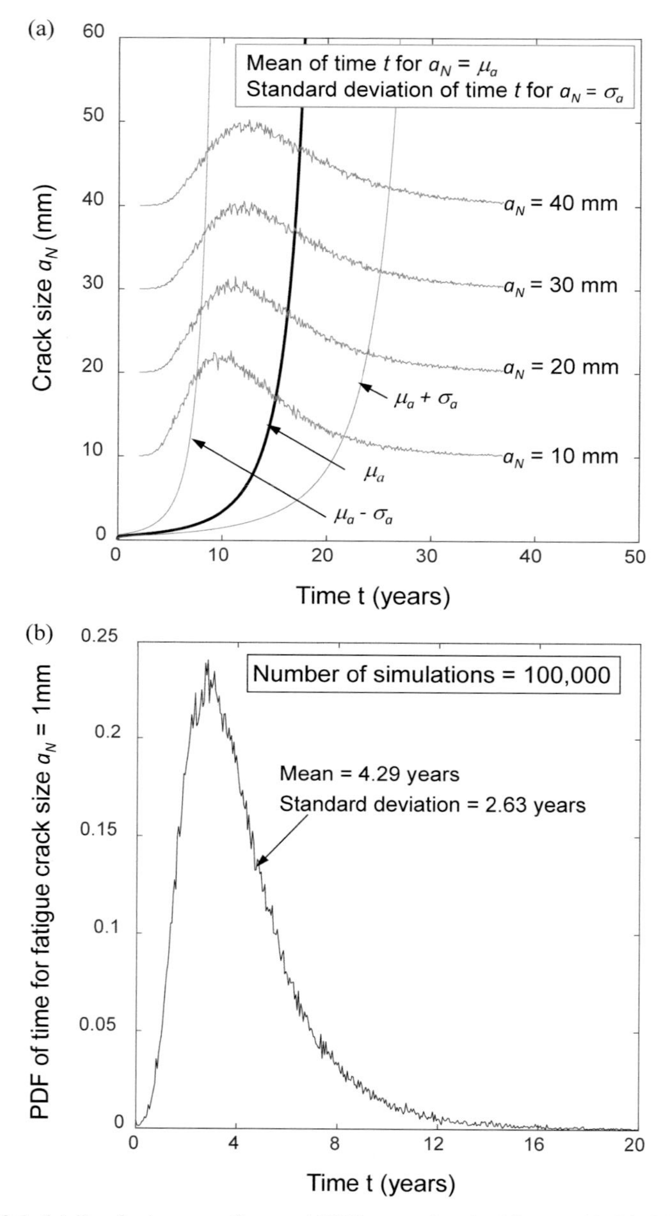

Figure 3.1 (a) Crack size over time and PDFs associated with a_N = 10, 20, 30 and 40 mm; and (b) PDF of time for fatigue crack size a_N to reach 1 mm.

3.3 Probability of Fatigue Damage Detection

The probability of fatigue damage detection is the conditional probability that the fatigue crack is detected when a specific crack size exists (Chung et al. 2006). The quality of an inspection method is related to the probability of detection for a given crack size, and detectable minimum crack size. The relation between the probability of fatigue crack damage detection P_{insp} and crack size a is quantified using different forms as follows:

a) Shifted exponential form (Packman et al. 1969):

$$P_{insp} = 1 - exp\left(-\frac{a - a_{min}}{\lambda}\right) \quad \text{for } a > a_{min} \tag{3.7}$$

where a_{min} is the minimum detectable crack size. λ is the parameter for inspection quality, which is larger than zero. A decrease in the value of λ is associated with a higher quality of inspection.

b) Log-logistic form (Berens and Hovey 1981; Berens 1989)

$$P_{insp} = \frac{exp[\chi + \kappa\, ln(a)]}{1 + exp[\chi + \kappa\, ln(a)]} \tag{3.8}$$

The statistical parameters χ and κ are estimated using the maximum likelihood method for a specific inspection method (Chung et al. 2006).

c) Cumulative lognormal distribution form (Crawshaw and Chambers 1984; Kim et al. 2013)

$$P_{insp} = 1 - \Phi\left(\frac{ln(a) - \alpha}{\beta}\right) \tag{3.9}$$

where $\Phi(\cdot)$ denotes the standard normal cumulative distribution function (CDF); α and β are the location and scale parameters, respectively.

d) Normal CDF form (Frangopol et al. 1997)

$$P_{insp} = \Phi\left(\frac{\delta - \delta_{0.5}}{\sigma_{\delta}}\right) \tag{3.10}$$

where δ = fatigue damage intensity; $\delta_{0.5}$ = fatigue damage intensity when an inspection method has a probability of damage detection of 0.5; and

σ_δ = standard deviation of $\delta_{0.5}$. The quality of the inspection method is represented by $\delta_{0.5}$. A smaller value of $\delta_{0.5}$ is associated with higher quality of inspection. The fatigue damage intensity in Eq. (3.10) is expressed as (Kim and Frangopol 2011d)

$$\delta = 0 \qquad \text{for } a < a_{min} \tag{3.11a}$$

$$\delta = \frac{a - a_{min}}{a_{max} - a_{min}} \qquad \text{for } a_{min} \le a < a_{max} \tag{3.11b}$$

$$\delta = 1 \qquad \text{for } a \ge a_{max} \tag{3.11c}$$

where a_{min} and a_{max} are the minimum and maximum detectable crack sizes, respectively.

Figure 3.2(a) shows the relation between the probability of fatigue crack detection and the crack size when the normal CDF form of Eq. (3.10) is applied with a_{min} = 1 mm and a_{max} = 50 mm in Eq. (3.11). Based on the crack propagation associated with Figure 3.1(a), the expected probability of fatigue crack detection over time after the crack size reaching 1.0 mm is illustrated in Figure 3.2(b). From Figure 3.2, it can be seen that an inspection with $\delta_{0.5}$ = 0.01 results in a higher P_{insp} than those with $\delta_{0.5}$ = 0.03 and 0.05, and $E(P_{insp}) = 0.5$ for $\delta_{0.5} = 0.01$ leads to the least crack size compared to $E(P_{insp}) = 0.5$ for $\delta_{0.5} = 0.03$ and $\delta_{0.5} = 0.05$.

By taking into account the relation among the inspection time, service life and probability of damage detection, the lifetime probability of fatigue crack damage detection P_{det} for N_{insp} inspections is formulated as (Soliman et al. 2013a)

$$P_{det} = \sum_{i=1}^{N_{insp}} \left[\prod_{j=1}^{i} \left\{ P\left(t_{insp,j} \le t_{life}\right) \cdot \left(1 - P_{insp,j-1}\right) \right\} \cdot P_{insp,i} \right] \tag{3.12}$$

where $P_{insp,i}$ is the probability of fatigue crack damage detection at the time of the *i*th inspection $t_{insp,i}$, t_{life} is the service life of the structure, and $P(t_{insp,j} \le t_{life})$ is the probability that the *j*th inspection is applied before t_{life}. $P_{insp,0}$ for j = 1 is equal to zero. For example, when two inspections are applied at $t_{insp,1}$ and $t_{insp,2}$, the probability of damage detection associated with the first inspection is $P(t_{insp,1} \le t_{life}) \cdot P_{insp,1}$, and the probability that the

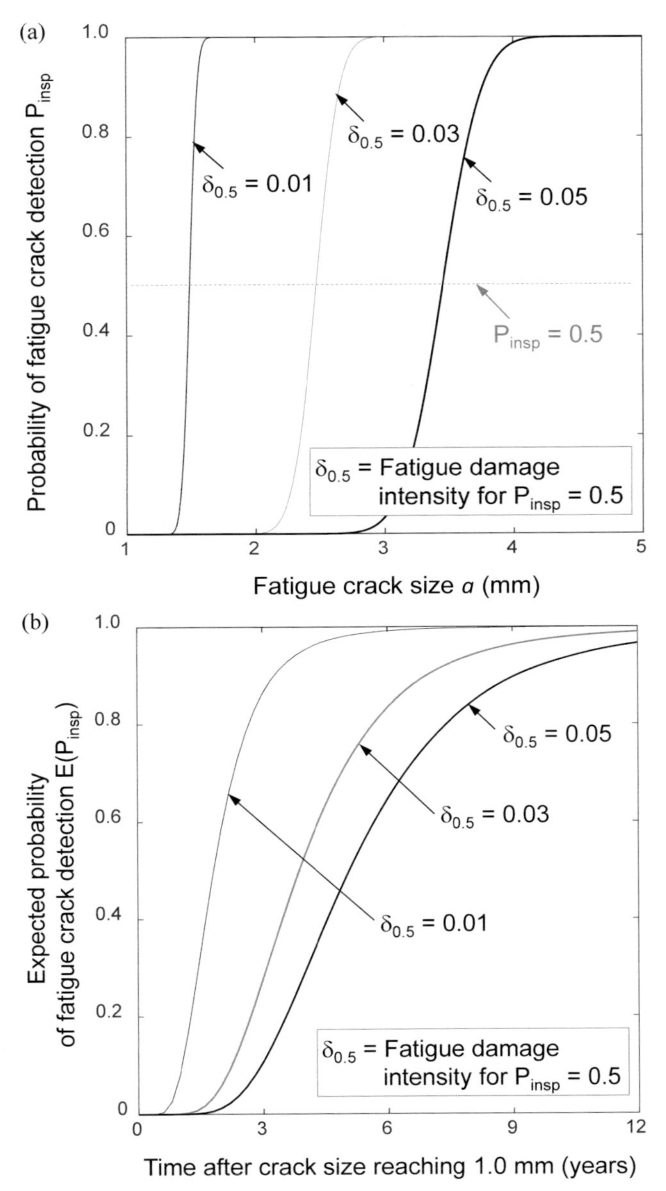

Figure 3.2 Probability of fatigue damage detection: (a) relation between probability of fatigue crack detection and crack size based on normal CDF form; (b) time-dependent expected probability of fatigue crack detection.

damage is detected at the second inspection is $P(t_{insp,1} \leq t_{life})\cdot P(t_{insp,2} \leq t_{life})$ $(1 - P_{insp,1})\cdot P_{insp,2}$. Finally, the lifetime probability of damage detection for two inspections becomes $P(t_{insp,1} \leq t_{life})\cdot P_{insp,1} + P(t_{insp,1} \leq t_{life})\cdot P(t_{insp,2} \leq t_{life})$ $(1 - P_{insp,1})\cdot P_{insp,2}$.

3.4 Damage Detection Time Under Uncertainty

In general, a fatigue crack propagates rapidly after a certain time, and failure occurs in a short time interval (Fatemi and Yang 1998; Fisher 1984; Fisher et al. 1998; Mohanty et al. 2009; Schijve 2003; Papazian et al. 2007). For this reason, fatigue cracks have to be detected as early as possible, and appropriate and timely maintenance actions should be applied to prevent unexpected structural failure (Glen et al. 2000; Dexter et al. 2003). However, due to the uncertainties associated with the prediction of fatigue crack propagation and fatigue crack detection, fatigue crack damage detection time cannot be predicted accurately, and, therefore, has to be predicted by using probabilistic concepts and methods (Kwon and Frangopol 2010, 2011, 2012; Kwon et al. 2013; Guedes Soares and Garbatov 1996a, 1996b; Soliman et al. 2015).

3.4.1 *Damage Detection Time and Delay for Inspection*

As illustrated in Figure 3.3, damage detection delay t_{del} is the time interval between the damage occurrence time t_{oc} and damage detection time t_{det}, which is estimated as

$$t_{del} = t_{det} - t_{oc} \qquad \text{for } t_{det} \geq t_{oc} \tag{3.13}$$

The damage detection delay is formulated by using the event tree model which considers all the possible cases and the corresponding consequences as shown in Figure 3.4.

For example, if two inspections at times $t_{insp,1}$ and $t_{insp,2}$ are performed to detect fatigue damage, there will be three possible cases as follows: (a)

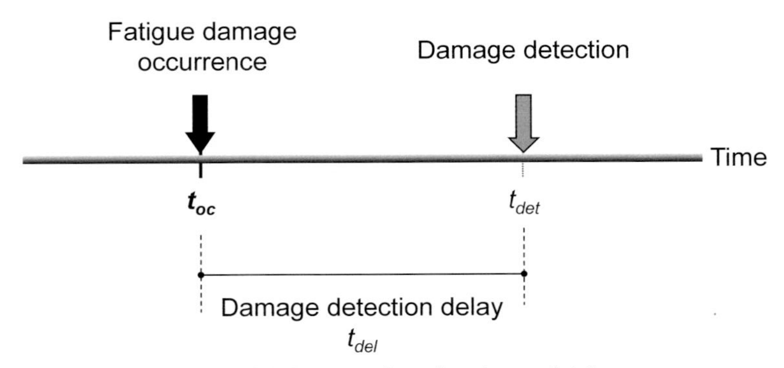

Figure 3.3 Damage detection time and delay.

case 1: $t_s \leq t_{oc} < t_{insp,1}$; (b) case 2: $t_{insp,1} \leq t_{oc} < t_{insp,2}$; (c) case 3: $t_{insp,2} \leq t_{oc} < t_e$. The gray circle at every inspection is a chance node in which two mutually exclusive events (i.e., detection and no detection) exist. t_s and t_e are the lower and upper bounds of damage occurrence, respectively. For case 1 (see Figure 3.4(a)), there will be three branches. If the damage is detected at the first inspection time $t_{insp,1}$, the damage detection delay will be $t_{insp,1} - t_{oc}$, and the associated branch probability is $P_{insp,1}$. The damage detection at the second inspection results in the damage detection delay $t_{insp,2} - t_{oc}$ and branch probability $(1 - P_{insp,1})\cdot P_{insp,2}$. If the damage is not detected until the second inspection, the associated damage detection delay and branch probability will be $t_e - t_{oc}$ and $(1 - P_{insp,1})\cdot(1 - P_{insp,2})$, respectively. Therefore, the damage detection delay for case 1 in Figure 3.4(a) can be estimated as

$$t_{del,case1} = P_{insp,1}\cdot(t_{insp,1} - t_{oc}) + (1 - P_{insp,1})\cdot P_{insp,2}\cdot(t_{insp,2} - t_{oc}) + (1 - P_{insp,1})\cdot(1 - P_{insp,2})\cdot(t_e - t_{oc}) \quad \text{for } t_s \leq t_{oc} < t_{insp,1} \tag{3.14}$$

In a similar way, the damage detection delay for cases 2 and 3 (see Figures 3.4(b) and 3.4(c)) is formulated as

$$t_{del,case2} = P_{insp,2}\cdot(t_{insp,2} - t_{oc}) + (1 - P_{insp,2})\cdot(t_e - t_{oc}) \quad \text{for } t_{insp,1} \leq t_{oc} < t_{insp,2} \tag{3.15a}$$

$$t_{del,case3} = t_e - t_{oc} \quad \text{for } t_{insp,2} \leq t_{oc} < t_e \tag{3.15b}$$

(a) **Case 1: $t_s \leq t_{oc} < t_{insp,1}$**

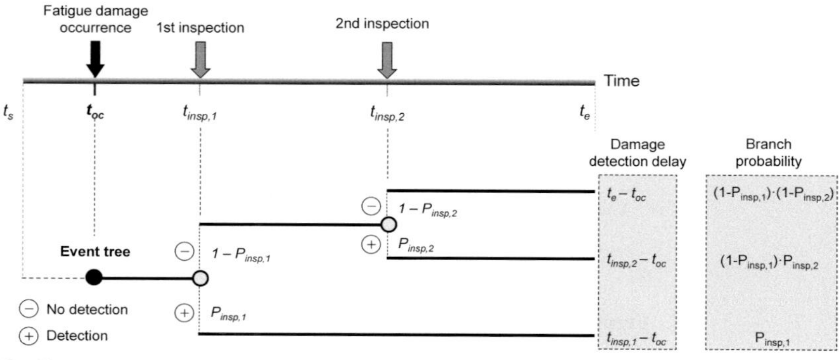

(b) **Case 2: $t_{insp,1} \leq t_{oc} < t_{insp,2}$**

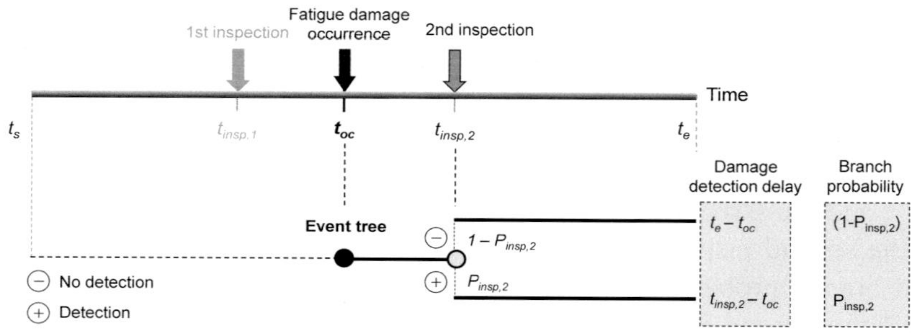

(c) **Case 3: $t_{insp,2} \leq t_{oc} < t_{insp,2}$**

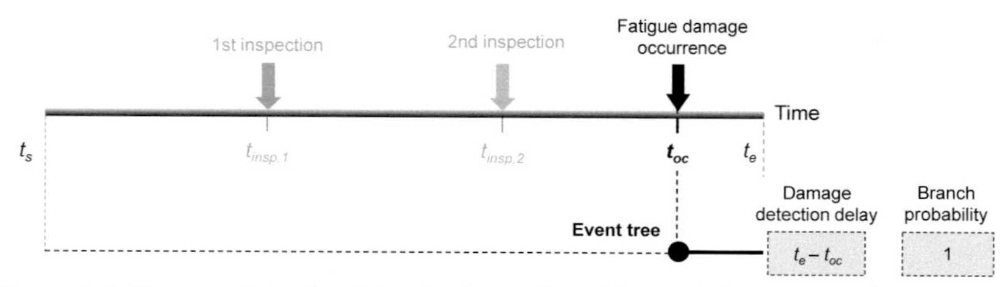

Figure 3.4 Damage detection delay for inspection: (a) case 1 ($t_s \leq t_{oc} < t_{insp,1}$); (b) case 2 ($t_{insp,1} \leq t_{oc} < t_{insp,2}$); (c) case 3 ($t_{insp,2} \leq t_{oc} < t_e$).

The lower and upper bounds of damage occurrence time (i.e., t_s and t_e in Figure 3.4) are defined as (Kim and Frangopol 2011c, 2011d)

$$t_s = F_T^{-1}(\Phi(-q)) \tag{3.16a}$$

$$t_e = F_T^{-1}(\Phi(q)) \tag{3.16b}$$

where $F_T^{-1}(\cdot)$ = the inverse CDF of the damage occurrence time t_{oc}, and q = scale parameter representing the time interval between lower and upper bounds of damage occurrence time. For example, if the damage occurrence time t_{oc} is normally distributed with the mean of 10 years and standard deviation of 3 years, and the scale parameter $q = 3$ is applied, the lower and upper bound of damage occurrence time t_s and t_e will be 1 and 19 years, respectively.

When N_{insp} inspections are applied at times $t_{insp,1}$, $t_{insp,2}$,...and $t_{insp,Ninsp}$, and damage occurs in the time interval between $t_{insp,i-1}$ and $t_{insp,i}$ (i.e., $t_{insp,i-1} \leq t_{oc} < t_{insp,i}$), the damage detection delay $t_{del,i}$ is formulated as

$$t_{del,i} = \sum_{k=i}^{N_{insp}+1} \left\{ \left(\prod_{j=1}^{k} \left(1 - P_{insp,j-1}\right) \right) \cdot P_{insp,k} \cdot t_{insp,k} \right\} - t_{oc} \quad \text{for } t_{insp,i-1} \leq t_{oc} < t_{insp,i} \tag{3.17}$$

where $P_{insp,0} = 0$ for $j = 1$, and $P_{insp,Ninsp+1} = 1$ for $k = N_{insp+1}$. As defined in Eq. (3.13) and Figure 3.3, the damage detection time $t_{det,i}$ for $t_{insp,i-1} \leq t_{oc} < t_{insp,i}$ is

$$t_{det,i} = \sum_{k=i}^{N_{insp}+1} \left\{ \left(\prod_{j=1}^{k} \left(1 - P_{insp,j-1}\right) \right) \cdot P_{insp,k} \cdot t_{insp,k} \right\} \quad \text{for } t_{insp,i-1} \leq t_{oc} < t_{insp,i} \tag{3.18}$$

The expected damage detection delay $E(t_{del})$ and expected damage detection time $E(t_{det})$ are formulated by treating the damage occurrence time t_{oc} as a random variable (Kim and Frangopol 2011c, 2011d) as

$$E\left(t_{del}\right) = \sum_{i=1}^{N_{insp}+1} \left\{ \int_{t_{insp,i-1}}^{t_{insp,i}} \left[t_{del,i} \cdot f_T\left(t_{oc}\right) \right] dt_{oc} \right\} \tag{3.19a}$$

$$E\left(t_{det}\right) = \sum_{i=1}^{N_{insp}+1} \left\{ \int_{t_{insp,i-1}}^{t_{insp,i}} \left[t_{det,i} \cdot f_T\left(t_{oc}\right) \right] dt_{oc} \right\} \tag{3.19b}$$

where $f_T(t_{oc})$ = PDF of the damage occurrence time t_{oc}.

3.4.2 *Damage Detection Time and Delay for Monitoring*

Appropriate determination of the locations to be monitored and installation of sensors can lead to accurate and reliable monitoring data acquisition, and increase of the probability of damage detection. If the damage is detected without delay during the monitoring, the expected damage detection delay $E(t_{del})$ and expected damage detection time $E(t_{det})$ can be formulated based on Eqs. (3.18) and (3.19) as (Kim and Frangopol 2011c, 2012)

$$E\left(t_{del}\right) = \sum_{i=1}^{N_{mon}+1} \left\{ \int_{t_{ms,i-1}+t_{md}}^{t_{ms,i}} \left[\left(t_{ms,i} - t_{oc}\right) \cdot f_T\left(t_{oc}\right) \right] dt_{oc} \right\} \tag{3.20a}$$

$$E\left(t_{det}\right) = \sum_{i=1}^{N_{mon}+1} \left\{ \int_{t_{ms,i-1}+t_{md}}^{t_{ms,i}} \left[t_{ms,i} \cdot f_T\left(t_{oc}\right) \right] dt_{oc} + \int_{t_{ms,i}}^{t_{ms,i}+t_{md}} \left[t_{oc} \cdot f_T\left(t_{oc}\right) \right] dt_{oc} \right\} \tag{3.20b}$$

where $t_{ms,i}$ = ith monitoring starting time; and t_{md} = monitoring duration. $t_{ms,i-1} + t_{md}$ for $i = 1$ and $t_{ms,i}$ for $N_{mon} + 1$ are t_s and t_e in Eq. (3.16), respectively.

3.4.3 *Damage Detection Time and Delay for Combined Inspection and Monitoring*

When combined inspection and monitoring is used to detect damage, the formulations of the expected damage detection delay $E(t_{del})$ and expected damage detection time $E(t_{det})$ are based on Eqs. (3.19) and (3.20). Assuming that one inspection and one monitoring are available, and the monitoring is applied after inspection, there will be four cases as shown in Figure 3.5: (a) case 1: $t_s \leq t_{oc} < t_{insp,1}$; (b) case 2: $t_{insp,1} \leq t_{oc} < t_{ms,1}$; (c) case 3: $t_{ms,1} \leq t_{oc} < t_{ms,1} + t_{md}$; (d) case 4: $t_{ms,1} + t_{md} \leq t_{oc} < t_e$. As a result, $E(t_{del})$ and $E(t_{det})$ are expressed as (Kim and Frangopol 2012)

$$E\left(t_{del}\right) = \int_{t_s}^{t_{insp,1}} \left[P_{insp,1} \cdot \left(t_{insp,1} - t_{oc}\right) + \left(1 - P_{insp,1}\right) \cdot \left(t_{ms,1} - t_{oc}\right) \right] \cdot f_T\left(t_{oc}\right) dt_{oc}$$
$$+ \int_{t_{insp,1}}^{t_{ms,1}} \left(t_{ms,1} - t_{oc}\right) \cdot f_T\left(t_{oc}\right) dt_{oc} + \int_{t_{ms,1}+t_{md}}^{t_e} \left(t_e - t_{oc}\right) \cdot f_T\left(t_{oc}\right) dt_{oc} \tag{3.21a}$$

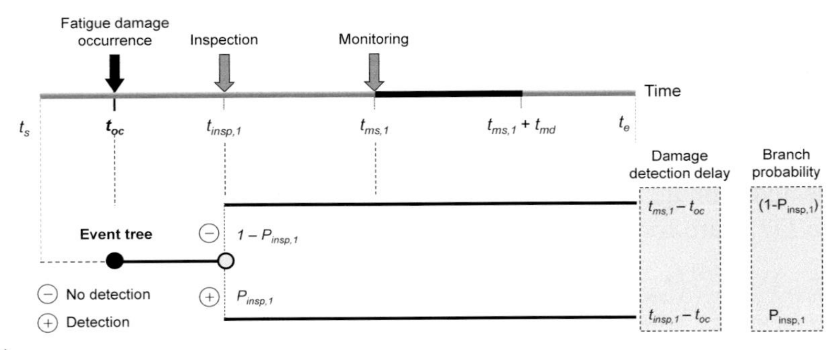

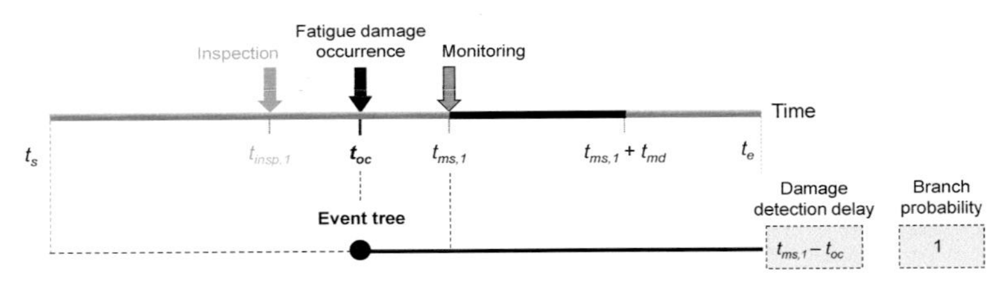

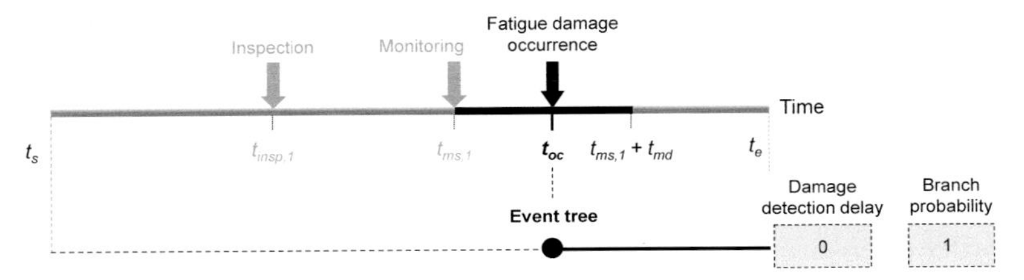

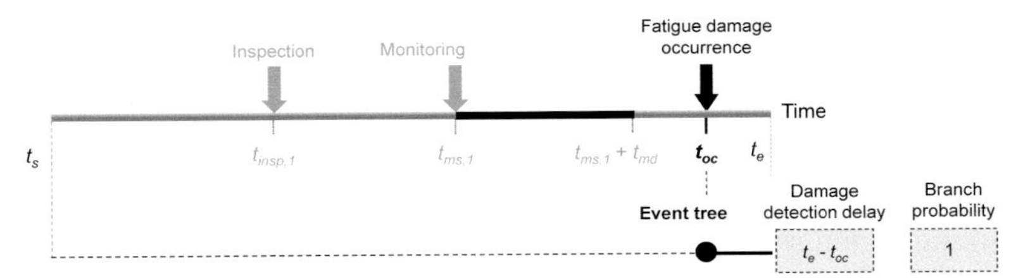

Figure 3.5 Damage detection delay for combined inspection and monitoring: (a) case 1: $t_s \leq t_{oc} < t_{insp,1}$; (b) case 2: $t_{insp,1} \leq t_{oc} < t_{ms,1}$; (c) case 3: $t_{ms,1} \leq t_{oc} < t_{ms,1} + t_{md}$; (d) case 4: $t_{ms,1} + t_{md} \leq t_{oc} < t_e$.

$$E\left(t_{det}\right)=\int_{t_s}^{t_{insp,1}}\left[P_{insp,1}\cdot t_{insp,1}+\left(1-P_{insp,1}\right)\cdot t_{ms,1}\right]\cdot f_T\left(t_{oc}\right)dt_{oc}$$
$$+\int_{t_{insp,1}}^{t_{ms,1}}t_{ms,1}\cdot f_T\left(t_{oc}\right)dt_{oc}+\int_{t_{ms,1}}^{t_{ms,1}+t_{md}}t_{oc}\cdot f_T\left(t_{oc}\right)dt_{oc}+\int_{t_{ms,1}+t_{md}}^{t_e}t_e\cdot f_T\left(t_{oc}\right)dt_{oc} \tag{3.21b}$$

Similarly, when the monitoring is performed before inspection, the expected damage detection delay $E(t_{del})$ and expected damage detection time $E(t_{det})$ are formulated as

$$E\left(t_{del}\right)=\int_{t_s}^{t_{ms,1}}\left(t_{ms,1}-t_{oc}\right)\cdot f_T\left(t_{oc}\right)dt_{oc}$$
$$+\int_{t_{ms,1}+t_{md}}^{t_{insp,1}}\left[P_{insp,1}\cdot\left(t_{insp,1}-t_{oc}\right)+\left(1-P_{insp,1}\right)\cdot\left(t_e-t_{oc}\right)\right]\cdot f_T\left(t_{oc}\right)dt_{oc}$$
$$+\int_{t_{insp,1}}^{t_e}\left(t_e-t_{oc}\right)\cdot f_T\left(t_{oc}\right)dt_{oc} \tag{3.22a}$$

$$E\left(t_{det}\right)=\int_{t_s}^{t_{ms,1}}t_{ms,1}\cdot f_T\left(t_{oc}\right)dt_{oc}+\int_{t_{ms,1}}^{t_{ms,1}+t_{md}}t_{oc}\cdot f_T\left(t_{oc}\right)dt_{oc}$$
$$+\int_{t_{ms,1}+t_{md}}^{t_{insp,1}}\left[P_{insp,1}\cdot t_{insp,1}+\left(1-P_{insp,1}\right)\cdot t_e\right]\cdot f_T\left(t_{oc}\right)dt_{oc}+\int_{t_{insp,1}}^{t_e}t\cdot f_T\left(t_{oc}\right)dt_{oc} \tag{3.22b}$$

3.4.4 Damage Detection Time-Based Probability of Failure

In order to ensure the structural safety, any damage should be detected and repaired before reaching the critical state. Accordingly, the safety margin can be defined in terms of crack size and time as

$$a_{mar}=a_{crt}-a_{oc} \tag{3.23a}$$

$$t_{mar}=t_{crt}-t_{oc} \tag{3.23b}$$

where a_{mar} = crack size-based safety margin; a_{crt} = critical crack size resulting in the structural failure; a_{oc} = crack size for the fatigue damage occurrence. t_{mar} is the time-based safety margin, and t_{crt} and t_{oc} are the times associated with the crack size a_{crt} and a_0, respectively, as shown in Figure 3.6. By considering the uncertainty associated with fatigue crack

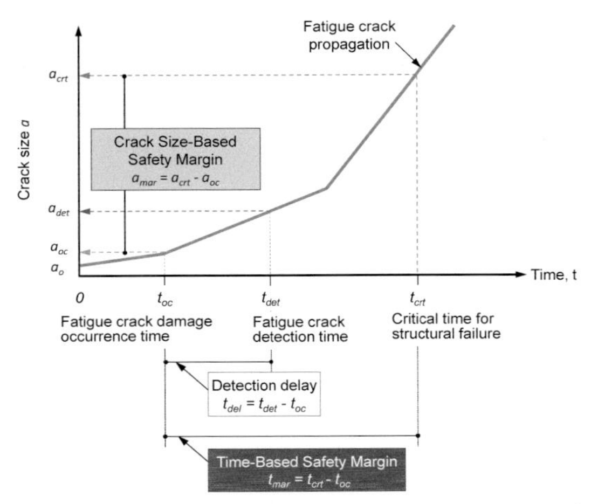

Figure 3.6 Crack size-based and time-based safety margins.

propagation over time, the PDFs of crack size based and time-based safety margin can be obtained.

Under the assumption that an appropriate and immediate maintenance action is applied after the damage is detected by inspection and/or monitoring, the time-based failure criterion can be expressed as (Kim and Frangopol 2011a)

$$t_{crt} - t_{det} = t_{mar} - t_{del} < 0 \tag{3.24}$$

where t_{det} is the damage detection time (Eq. (3.18)), and t_{del} is the damage detection delay (Eq. (3.17)). Considering the uncertainties associated with variables in Eq. (3.24) (i.e., t_{crt}, t_{det}, t_{mar}, and t_{del}), the damage detection time-based probability of failure is expressed as

$$P_f = P(t_{crt} - t_{det} < 0) = P(t_{mar} - t_{del} < 0) \tag{3.25}$$

Figure 3.7 provides the PDFs of the fatigue damage occurrence time t_{oc}, critical time for structural failure t_{crt}, and time-based safety margin t_{mar} defined in Eq. (3.23b). The fatigue crack propagation model associated with Figure 3.1 and Table 3.1 is assumed in Figure 3.7, in which the joint between the bottom plate and longitudinal plate of a ship hull is investigated. It is assumed that the crack size for the fatigue damage occurrence a_{oc} is

1 *mm* and the critical crack size a_{crt} is 20 *mm*. The difference between the expected critical time for failure $E(t_{crt})$ and the expected damage occurrence time $E(t_{oc})$ is equal to the expected time-based safety margin $E(t_{mar})$ (i.e., 15.49 years – 4.29 years = 11.20 years) as shown in Figure 3.7. The correlation coefficients among t_{oc}, t_{crt} and t_{mar} are presented in Table 3.2.

In order to detect the fatigue damage, various nondestructive inspection methods including eddy current (EC) technique, ultrasonic (UL) inspection, and liquid penetration (LP) inspection can be used (Kwon

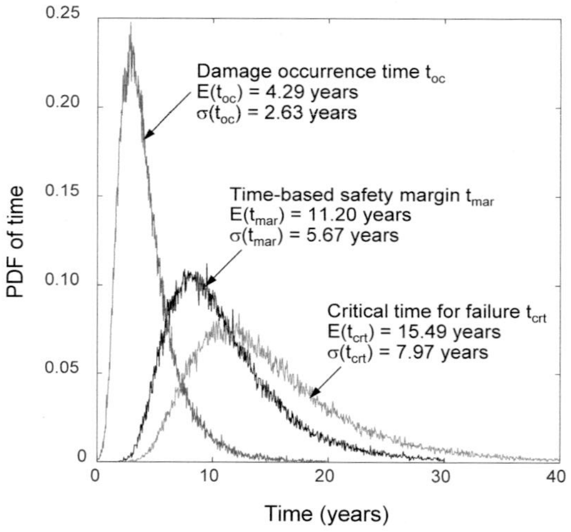

Figure 3.7 PDFs of fatigue crack damage initiation time t_{oc}, critical time for failure t_{crt}, and time-based safety margin t_{mar}.

Color version at the end of the book

Table 3.2 Coefficients of correlation among the fatigue damage occurrence time t_{oc}, critical time for failure t_{crt} and time-based safety margin t_{mar}.

	Fatigue damage occurrence time t_{oc}	**Critical time for failure t_{crt}**	**Time-based safety margin t_{mar}**
t_{oc}	1	0.917	0.828
t_{crt}	0.917	1	0.983
t_{mar}	0.828	0.983	1

and Frangopol 2011). Figure 3.8 shows the PDFs of the damage detection delay t_{del} and difference between time-based safety margin t_{mar} and damage detection delay t_{del} (i.e., $t_{mar} - t_{del}$), when the two-time UL inspections are

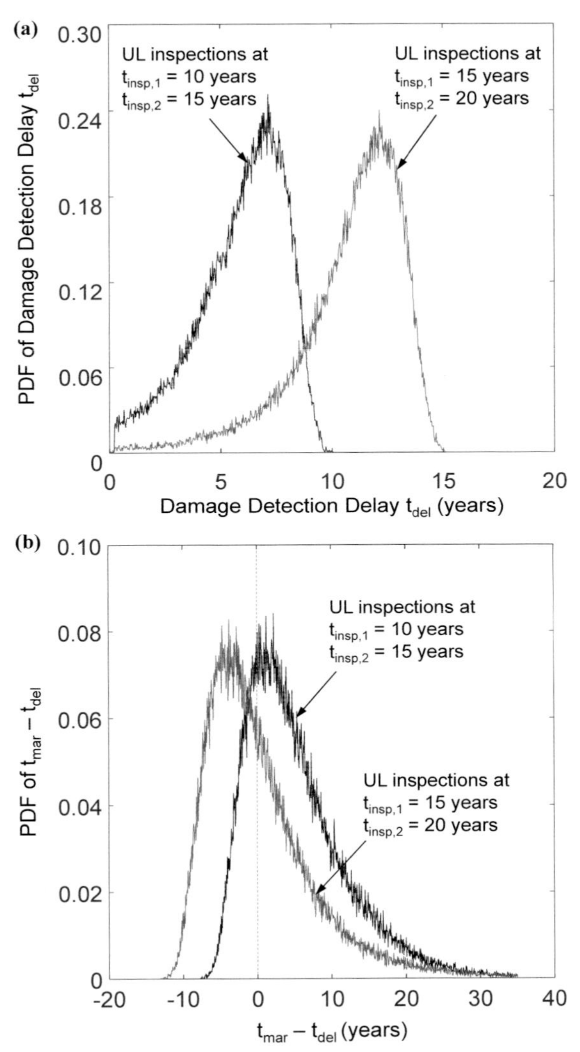

Figure 3.8 When two-time UL inspections are applied: (a) PDFs of damage detection delay t_{del}; (b) difference between time-based safety margin and damage detection delay $t_{mar} - t_{del}$.

applied to detect the fatigue crack damage. Herein, it is assumed that the probability of fatigue damage detection associated with the UL inspection is represented by the cumulative lognormal distribution form (see Eq. (3.9)) with the parameters $\alpha = 0.122$ and $\beta = -0.305$ (Forsyth and Fahr 1998). The UL inspections performed at $t_{insp,1} = 10$ years and $t_{insp,2} = 15$ years result in the lifetime probability of fatigue crack damage detection $P_{det} = 0.77$ and the expected damage detection delay $E(t_{del}) = 5.93$ years, as shown in Figure 3.8(a) and Table 3.3. If the UL inspections are applied at $t_{insp,1} = 15$ years and $t_{insp,2} = 20$ years instead of $t_{insp,1} = 10$ years and $t_{insp,2} = 15$ years, the lifetime probability of fatigue crack damage detection P_{det} decreases from 0.77 to 0.42, and the expected damage detection delay $E(t_{del})$ increases from 5.93 years to 10.81 years. Figure 3.8(b) shows the PDFs of $t_{mar} - t_{del}$, when the two-time UL inspections are applied at $t_{insp,1} =$ 10 years and $t_{insp,2} = 15$ years, and at $t_{insp,1} = 15$ years and $t_{insp,2} = 20$ years. It should be noted that the area under the PDF of $t_{mar} - t_{del}$ below zero is the damage detection time-based probability of failure P_f, as indicated in Eq. (3.25). The UL inspections at $t_{insp,1} = 10$ years and $t_{insp,2} = 15$ years lead to $P_f = 0.23$ (see Figure 3.8(b) and Table 3.3). By performing the UL inspections at $t_{insp,1} = 15$ years and $t_{insp,2} = 20$ years instead of $t_{insp,1} = 10$ years and $t_{insp,2} = 15$ years, P_f can increase from 0.23 to 0.58. From Figure 3.8 and Table 3.3, it can be seen that an increase in lifetime probability of fatigue crack damage detection P_{det} leads to a reduction of both $E(t_{del})$ and P_f.

Table 3.3 Lifetime probability of fatigue damage detection, expected damage detection delay and damage detection time-based probability of failure.

Ultrasonic inspection time (years)		Lifetime probability of fatigue damage detection P_{det}	Expected damage detection delay $E(t_{del})$ (years)	Damage detection time-based probability of failure P_f
$t_{insp,1}$	$t_{insp,2}$			
5	10	0.985	2.577	0.0076
10	15	0.765	5.932	0.233
15	20	0.418	10.805	0.582

3.5 Conclusions

The uncertainties associated with fatigue crack propagation and damage detection can result in damage detection delay and unexpected structural failure. In order to ensure the structural safety during the service life of a fatigue-sensitive structure, and manage the service life effectively and efficiently under uncertainty, probabilistic concepts and methods need to be applied in a rational way. This chapter uses these concepts and methods for service life management of fatigue-sensitive structures by presenting an overview of the time-dependent fatigue crack propagation under uncertainty, and evaluating the (a) probability of fatigue damage detection for single and multiple inspections, (b) damage detection time and delay when inspections and/or monitorings are performed, and (c) damage detection time-based probability of failure. The effects of inspection times on the probability of fatigue damage detection, damage detection delay and damage detection time-based probability of failure are investigated. The results provided in this chapter are used for optimum inspection and monitoring planning of fatigue-sensitive structures in the next chapter.

Chapter 4

Damage Detection Based Optimum Inspection and Monitoring Planning

CONTENTS

ABSTRACT

Optimum inspection, monitoring and maintenance strategies should be established by using an appropriate optimization process to maximize the efficiency and effectiveness of service life management. Chapter 4 focuses on such an optimization process using the probabilistic fatigue crack damage detection-based objectives. The objectives used in this chapter include maximizing the lifetime probability of fatigue crack damage detection, minimizing the expected fatigue crack damage detection delay, and minimizing the fatigue crack damage detection time–based probability of failure. By solving the single–objective optimization problem for inspection planning, the inspection times are estimated, and the effects of the number of inspections and the inspection quality on optimum inspection planning are investigated. The single-objective optimization for optimum monitoring planning produces optimum monitoring starting times for the given number of monitorings and monitoring durations. The relation among the number of monitorings, monitoring duration and optimum monitoring planning is studied. Furthermore, the optimum combined inspection–monitoring planning is also presented.

4.1 Introduction

The primary purpose of the service life management of fatigue-sensitive structures consists in the efficient and effective structural safety improvement and service life extension through appropriate and timely inspection, monitoring and maintenance actions (Frangopol and Kim 2014b; NCHRP 2003; van Noortwijk and Frangopol 2004; Zayed et al. 2002). In order to maximize the efficiency and effectiveness of service life management, optimum inspection, monitoring and maintenance strategies should be established through an optimization process (Barone and Frangopol 2013; Barone et al. 2014; Dong and Frangopol 2015; Frangopol and Maute 2003; Kong and Frangopol 2004; Lukic and Cremona 2001; Miyamoto et al. 2000). For this reason, during the past decade, the optimization techniques have been extensively used for the service life management of engineering structures such as aerospace structures, ships and bridges (Frangopol 2011; Frangopol et al. 2012; Frangopol and Soliman 2016; Sánchez-Silva et al. 2016). In general, the formulation of the optimization problem consists of objectives, design variables, constraints and given conditions (Arora 2016). The objective functions for the optimum service life management of fatigue-sensitive structures should take into account fatigue crack growth and damage detection under uncertainty (Kim et al. 2013).

In this chapter, the optimum inspection and monitoring planning for service life management of fatigue-sensitive structures is addressed using probabilistic fatigue crack damage detection-based objectives. The objectives include maximizing the lifetime probability of fatigue crack damage detection, minimizing the expected fatigue crack damage detection delay, and minimizing the fatigue crack damage detection time-based probability of failure. The probabilistic concepts and formulations related to these objective functions are provided in Chapter 3. Through the single-objective optimization process for inspection planning, the inspection times are estimated, and the effects of number of inspections and inspection quality on the value of objectives are investigated. The single-objective optimization for optimum monitoring planning produces an optimum monitoring starting time for given number of monitorings and monitoring durations. The relationship among the number of monitorings, monitoring duration and objective values is provided. Furthermore, the optimum combined inspection and monitoring planning is presented.

4.2 Optimum Inspection Planning

Optimum inspection planning can be obtained as a solution of a single-objective optimization problem, considering one of the following three objectives $O_{I,1}$ = maximizing the lifetime probability of fatigue crack damage detection P_{det}, $O_{I,2}$ = minimizing the expected fatigue crack damage detection delay $E(t_{del})$, and $O_{I,3}$ = minimizing the fatigue crack damage detection time-based probability of failure P_f. The formulation of P_{det}, $E(t_{del})$ and P_f can be found in Chapter 3 (see Eqs. (3.12), (3.19) and (3.25), respectively). If the total number of inspections and the type of the inspection method are given, the design variable of the optimization for inspection planning is the inspection time.

The optimum inspection plannings based on $O_{I,1}$, $O_{I,2}$ and $O_{I,3}$ are applied to the joint between the bottom plate and longitudinal plate of a ship hull subjected to fatigue. The fatigue crack propagation over time for this location is described in Chapter 3 (see Table 3.1 and Figure 3.1). The cumulative lognormal distribution function form of Eq. (3.9) is used to represent the three types of nondestructive inspection methods (i.e., eddy current (EC) technique, ultrasonic (UL) inspection, and liquid penetrant (LP) inspection). The location and scale parameters (i.e., α and β) of the cumulative lognormal distribution function form for these three inspection methods are provided in Table 4.1. The location and scale parameters should be determined using extensive experimental investigations and data, since the geometry and location of the crack, environmental conditions and inspector training can affect the determination of these parameters (Kwon and Frangopol 2011; Soliman et al. 2013a). Figure 4.1 compares the probabilities of fatigue crack detection P_{insp} for the three inspection types EC, UL and LP. As shown in Figure 4.1, the EC inspection results in the highest inspection quality (i.e., highest P_{insp}) among the three inspection types considered. The optimization

Table 4.1 Parameters of cumulative lognormal distribution function* for eddy current technique, ultrasonic inspection and liquid penetrant inspection (Forsyth and Fahr 1998).

Inspection methods	**Location parameter α**	**Scale parameter β**
Eddy current technique (EC)	–0.968	–0.571
Ultrasonic inspection (UL)	0.122	–0.305
Liquid penetrant inspection (LP)	0.829	–0.423

*Lognormal distribution function form: $P_{insp} = 1 - \Phi\left(\frac{ln(a) - \alpha}{\beta}\right)$.

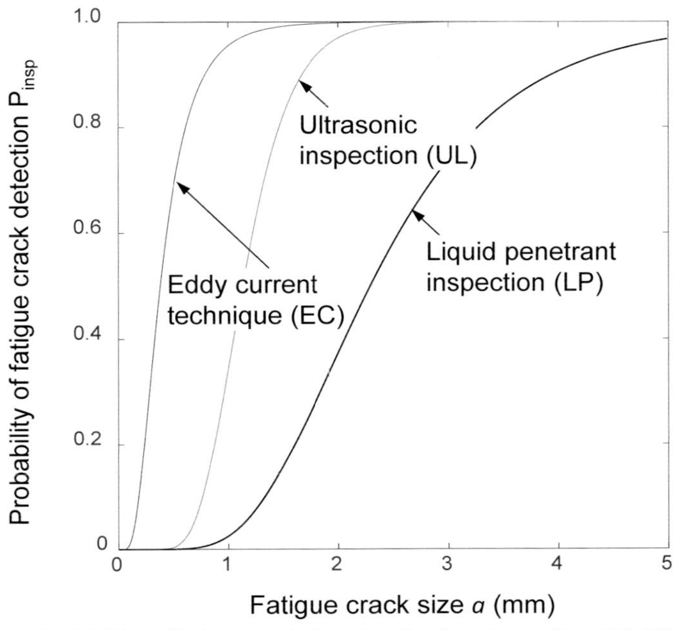

Figure 4.1 Probability of fatigue crack detection for three inspections EC, UL and LP.

problems in this chapter are solved using the constrained nonlinear minimization algorithm provided in MATLAB® version 2016b (MathWorks 2016). The genetic algorithm (GA) in MATLAB version 2016b is used to check if the optimum solution is global.

4.2.1 *Maximizing Lifetime Probability of Damage Detection*

Optimum inspection planning can be based on $O_{I,1}$ (i.e., maximizing the lifetime probability of fatigue crack damage detection P_{det}). The associated formulation is

Find $\mathbf{t_{insp}} = \{t_{insp,1}, t_{insp,2}, \ldots, t_{insp,Ninsp}\}$ (4.1a)

to minimize $$P_{det} = \sum_{i=1}^{N_{insp}} \left[\prod_{j=1}^{i} \left\{ P\left(t_{insp,j} \leq t_{life}\right) \cdot \left(1 - P_{insp,j-1}\right) \right\} \cdot P_{insp,i} \right] \quad (4.1b)$$

such that $t_{insp,i} - t_{insp,i-1} \geq 1$ year and $t_{insp,i} \leq 20$ years (4.1c)

given N_{insp} and type of the inspection method (4.1d)

where $\mathbf{t_{insp}}$ = vector of design variables (i.e., inspection times); $t_{insp,i}$ = ith inspection time (years); $P_{insp,i}$ = probability of fatigue crack damage detection at the time $t_{insp,i}$. The time interval between inspections has to be larger than 1 year, and the inspection time should be less than 20 years, as indicated in Eq. (4.1c). The formulation of the objective function P_{det} associated with the given total number of inspections N_{insp} is described in Chapter 3 (see Eq. (3.12)).

The optimum inspection times and corresponding lifetime probability of fatigue damage detection for three types of the nondestructive inspection methods are presented in Figure 4.2. The two UL inspections applied at $t_{insp,1}$ = 5.80 years and $t_{insp,2}$ = 11.94 years can lead to the maximum P_{det} = 0.929 (see Figure. 4.2(b)). If the number of inspections increases from two to three, the maximum expected P_{det} is 0.972 (see Figure 4.2(c)). When the three EC inspections instead of UL inspections are available, the EC inspections have to be performed at $t_{insp,1}$ = 2.87 years, $t_{insp,2}$ = 6.26 years and $t_{insp,3}$ = 16.18 years in order to maximize P_{det}. The relation between P_{det} and N_{insp} for the three inspection types considered EC, UL and LP is illustrated in Figure 4.3. From Figures 4.2 and 4.3, it can be seen that the EC inspections result in highest P_{det} among the three inspection types, and an increase in the number of inspections N_{insp} causes an increase in P_{det}, and a decrease in the first inspection time. It is significant to note that the single optimization problem to maximize P_{det} without comparing the service life and inspection time (i.e., $P(t_{insp,j} \leq t_{life})$ in Eq. (4.1b)) will result in the inspection times near its upper bound (i.e., 20 years as indicated in Eq. (4.1c)), since the fatigue crack propagates over time, and the probability of damage detection P_{insp} increases accordingly with crack growth.

4.2.2 Minimizing Expected Fatigue Crack Damage Detection Delay

The objective $O_{1,2}$ (i.e., minimizing the expected fatigue crack damage detection delay $E(t_{del})$) can be used to establish the optimum inspection plan. The single objective optimization for $O_{1,2}$ is formulated as

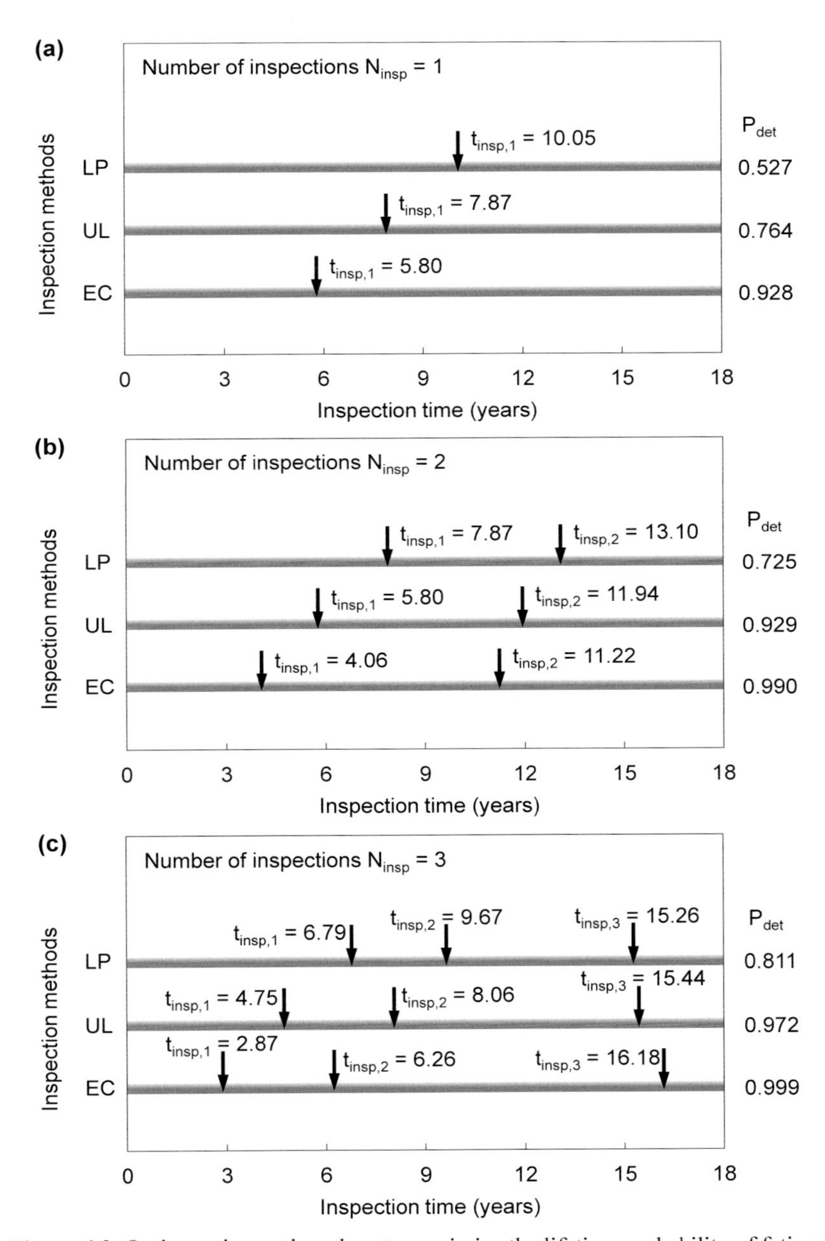

Figure 4.2 Optimum inspection plans to maximize the lifetime probability of fatigue crack damage detection P_{det}: (a) $N_{insp} = 1$; (b) $N_{insp} = 2$; (c) $N_{insp} = 3$.

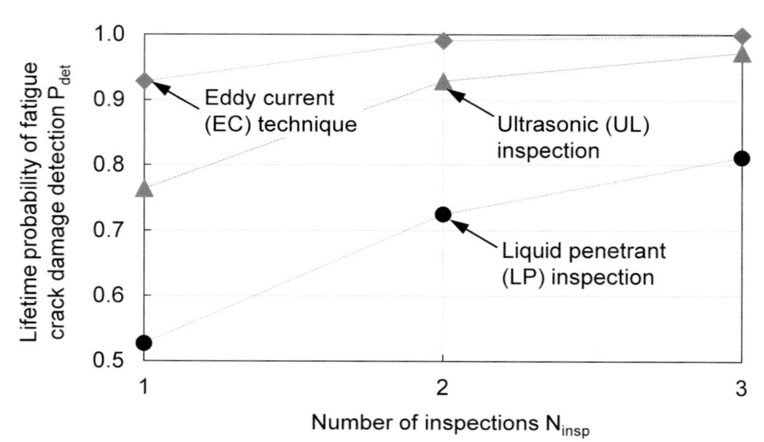

Figure 4.3 Relation between number of inspections N_{insp} and lifetime probability of fatigue crack damage detection P_{det} for three inspections EC, UL and LP.

$$\text{Find } \mathbf{t_{insp}} = \{t_{insp,1}, t_{insp,2}, \ldots, t_{insp,Ninsp}\} \tag{4.2a}$$

$$\text{to minimize } E(t_{del}) = \sum_{i=1}^{N_{insp}+1} \left\{ \int_{t_{insp,i-1}}^{t_{insp,i}} \left[t_{del,i} \cdot f_T(t_{oc}) \right] dt_{oc} \right\} \tag{4.2b}$$

where $\mathbf{t_{insp}}$ is the vector of the design variables (i.e., inspection times), and $t_{del,i}$ is the damage detection delay when damage occurs in the time interval between $t_{insp,i-1}$ and $t_{insp,i}$. $f_T(t_{oc})$ is the PDF of the damage occurrence time t_{oc}, which is provided in Figure 3.7. The detailed formulation of $t_{del,i}$ can be found in Chapter 3 (see Eq. (3.17)). The constraints and given conditions of this optimization are identical to the optimization problem associated with $O_{I,1}$ (see Eqs. (4.1c) and (4.1d)).

The optimum inspection plans for the three inspection types EC, UL and LP are illustrated in Figure 4.4. In order to minimize the expected damage detection delay $E(t_{del})$ with the two times EC inspections, they have to be performed at 4.52 years and 8.67 years. If the UL is applied instead of the EC, 5.44 years and 9.59 years will be the optimum inspection times, and $E(t_{del})$ will be equal to 3.56 years. Furthermore, when the number of UL inspections increases from two to three, $E(t_{del})$ can be decreased by 23.3% (from 3.56 years to 2.73 years). The effects of number of inspections N_{insp} and

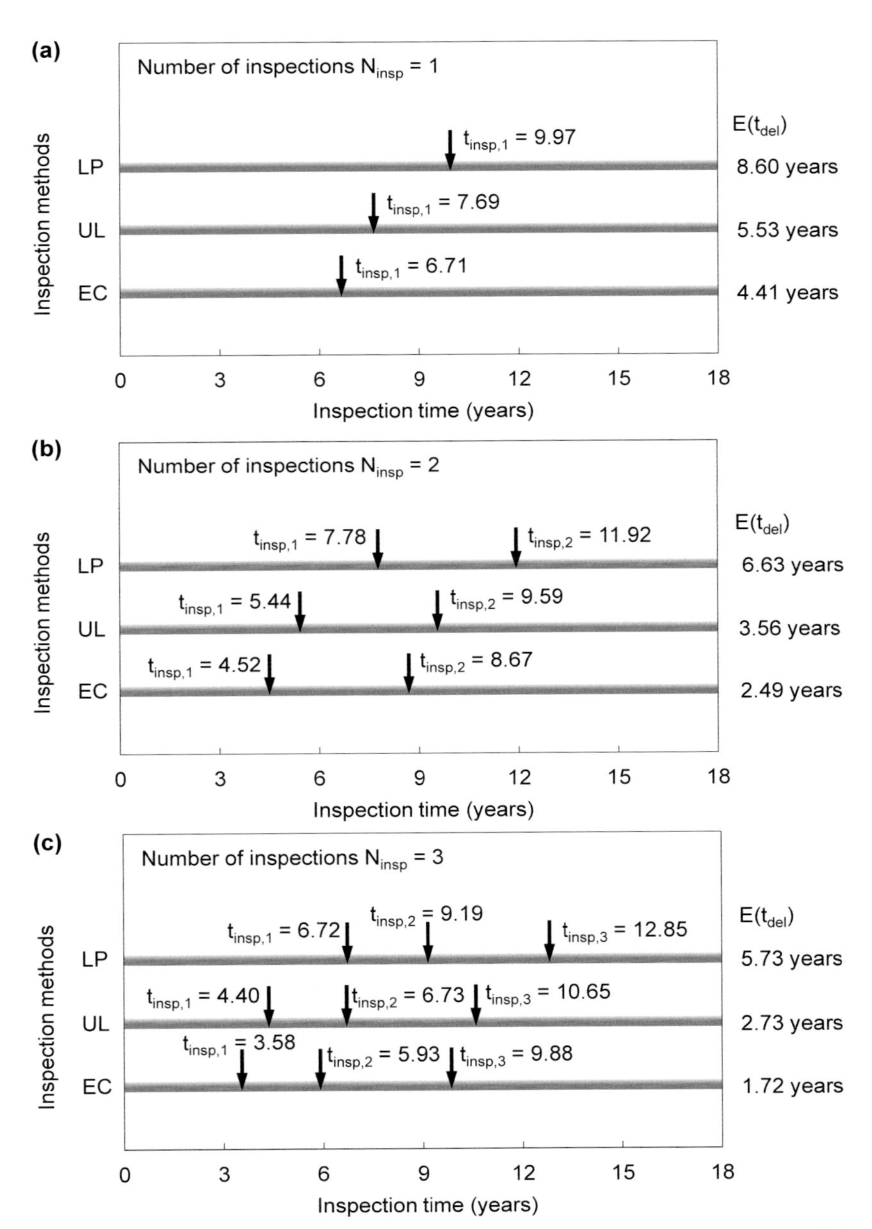

Figure 4.4 Optimum inspection plans to minimize the expected damage detection delay $E(t_{del})$: (a) $N_{insp} = 1$; (b) $N_{insp} = 2$; (c) $N_{insp} = 3$.

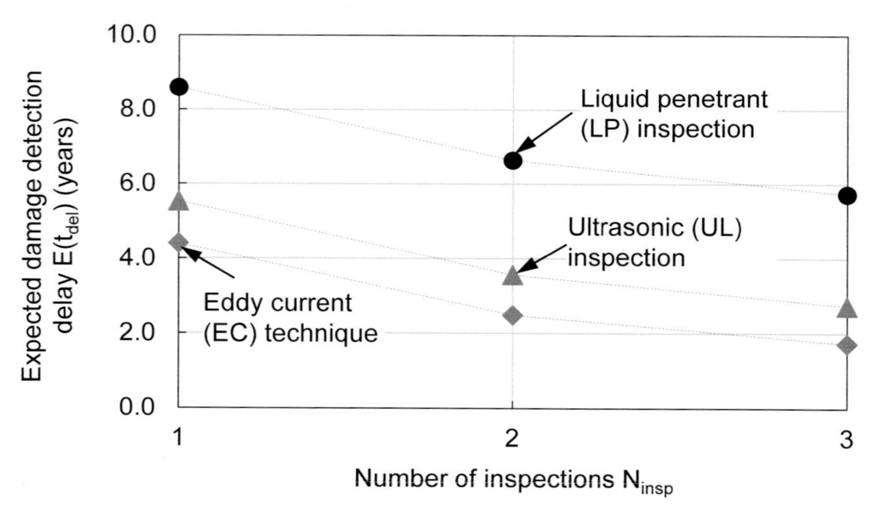

Figure 4.5 Relation between number of inspections N_{insp} and expected damage detection delay $E(t_{del})$ for three inspections EC, UL and LP.

type of an inspection method on the expected damage detection delay $E(t_{del})$ are presented in Figure 4.5. It can be concluded that the application of the EC associated with the highest inspection quality leads to the least $E(t_{del})$ among the three inspection types, and an increase in the number of inspection results in the reduction of $E(t_{del})$.

4.2.3 *Minimizing Damage Detection Time-Based Probability of Failure*

The formulation of the optimization problem associated with $O_{I,3}$ (i.e., minimizing the fatigue crack damage detection time-based probability of failure P_f) is

Find $\mathbf{t}_{\mathbf{insp}} = \{t_{insp,1}, t_{insp,2}, \ldots, t_{insp,Ninsp}\}$ (4.3a)

to minimize $P_f = P(t_{crt} - t_{det} < 0) = P(t_{mar} - t_{del} < 0)$ (4.3b) where t_{crt} = time for the critical crack size resulting in structural failure a_{cr} (20 *mm* herein); t_{det} = damage detection time; t_{mar} = time-based safety margin; t_{del} = damage detection delay. The design variables of this optimization problem are the inspection times $t_{insp,i}$, as indicated in Eq. (4.3a). The formulation of the fatigue crack damage detection time-based probability of failure P_f is described in Chapter 3 (see Eqs. (3.23), (3.24) and (3.25)). The PDFs of t_{crt} and t_{mar} in Figure 3.7 are used to formulate the objective function P_f. The times t_{crt} and t_{del} depend on the inspection times and methods as explained in Chapter 3. The constraints and given conditions of Eqs. (4.1c) and (4.1d) are also applied to this optimization problem.

The solutions of the optimization problem defined in Eq. (4.3) (i.e., inspection times) are provided in Figure 4.6. Figures 4.6(a), 4.6(b) and 4.6(c) are associated with the optimum inspection plans when the number of inspections N_{insp} is one, two and three, respectively. The effects of number of inspections and inspection methods (i.e., EC, UL and LP) on the fatigue damage detection time-based probability of failure P_f are presented in Figure 4.7. As shown in Figures 4.6 and 4.7, the EC inspection method results in the earliest inspection and the smallest probability of failure P_f among the three inspection methods. Conversely, the LP inspection is associated with the latest inspection and the largest P_f.

4.3 Optimum Monitoring Planning

The two objectives $O_{M,1}$ = minimizing the expected fatigue crack damage detection delay $E(t_{del})$ and $O_{M,2}$ = minimizing the fatigue crack damage detection time-based probability of failure P_f are used to establish the optimum monitoring plan. Chapter 3 explains how these two objective functions are formulated (see Eqs. (3.20a) and (3.25), respectively). Through the optimization process based on these two objectives, the optimum monitoring starting time can be obtained. In this section, the optimum monitoring planning is applied to the same fatigue-sensitive location (i.e., joint between bottom plate and longitudinal plate of a ship structure) used in the previous section (i.e., Section 4.2).

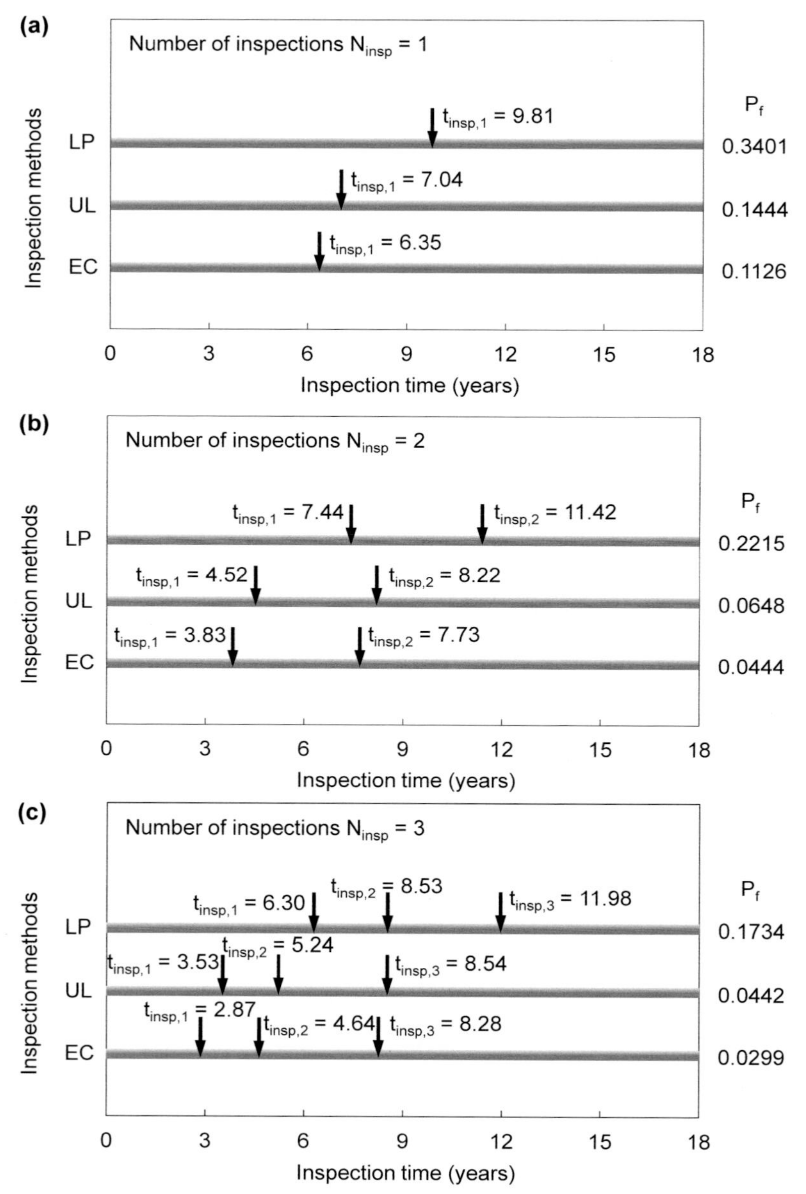

Figure 4.6 Optimum inspection plans to minimize the fatigue crack damage detection time-based probability of failure P_f: (a) $N_{insp} = 1$; (b) $N_{insp} = 2$; (c) $N_{insp} = 3$.

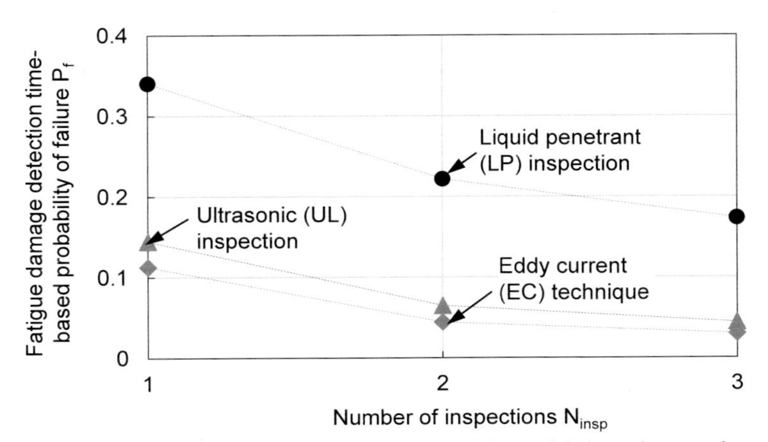

Figure 4.7 Relation between number of inspections N_{insp} and fatigue damage detection time-based probability of failure P_f for three inspections EC, UL and LP.

4.3.1 Minimizing Expected Fatigue Crack Damage Detection Delay

The optimum monitoring planning based on $O_{M,1}$ (i.e., minimizing the expected fatigue crack damage detection delay $E(t_{del})$) is formulated as

$$\text{Find } \mathbf{t_{ms}} = \{t_{ms,1}, t_{ms,2}, \ldots, t_{ms,Nmon}\} \tag{4.4a}$$

$$\text{to minimize } E(t_{del}) = \sum_{i=1}^{N_{mon}+1} \left\{ \int_{t_{ms,i-1}+t_{md}}^{t_{ms,i}} \left[\left(t_{ms,i} - t_{oc}\right) \cdot f_T\left(t_{oc}\right) \right] dt_{oc} \right\} \tag{4.4b}$$

$$\text{such that } 1 \text{ year} \leq t_{ms,i} - (t_{ms,i-1} + t_{md}) \leq 20 \text{ years} \tag{4.4c}$$

$$\text{given } N_{mon} \text{ and } t_{md} \tag{4.4d}$$

where $\mathbf{t_{ms}}$ = vector of design variables consisting of monitoring starting times $t_{ms,i}$ (years); N_{mon} = number of monitorings; t_{md} = monitoring duration (years); and $f_T(t_{oc})$ is the PDF of the damage occurrence time t_{oc} (see Figure 3.7). The non-monitoring time interval (i.e., $t_{ms,i} - (t_{ms,i-1} + t_{md})$) has to be larger than 1 year and less than 20 years, as indicated in Eq. (4.4c). The number of monitorings N_{mon} and monitoring duration t_{md} are given for this optimization problem.

The solutions (i.e., monitoring starting times $t_{ms,i}$) obtained from the optimization problem defined in Eq. (4.4) are provided in Figure 4.8. For

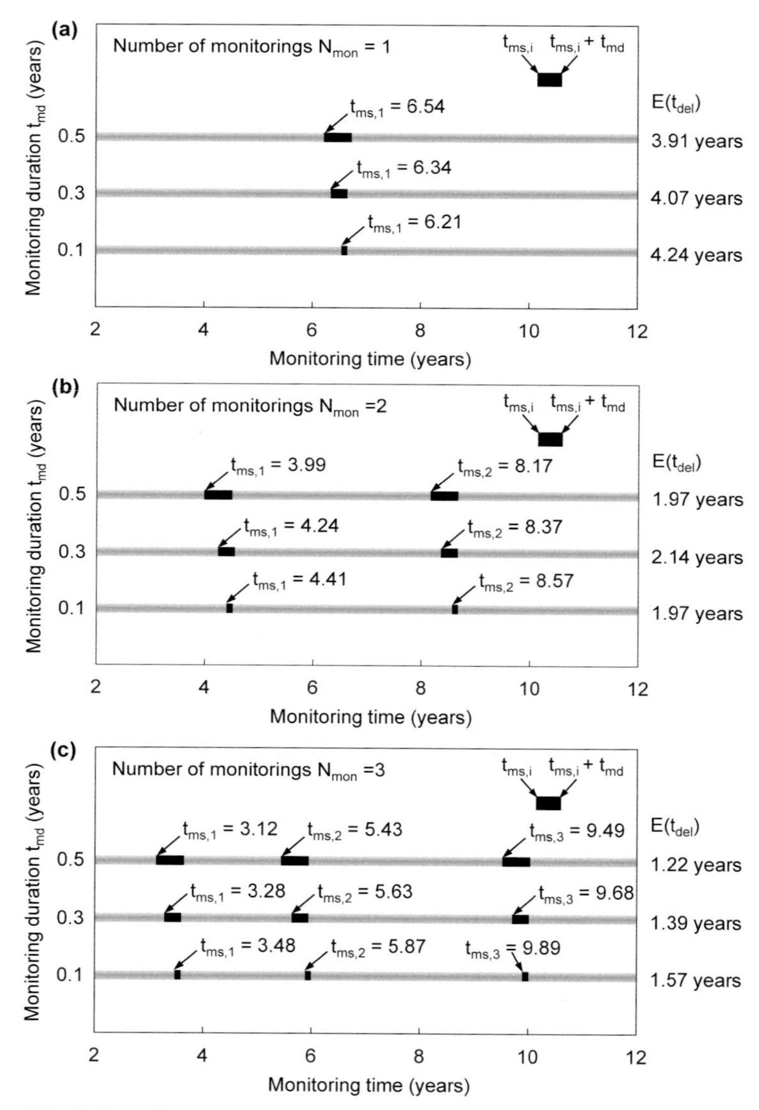

Figure 4.8 Optimum monitoring plans to minimize the expected damage detection delay $E(t_{del})$: (a) $N_{mon} = 1$; (b) $N_{mon} = 2$; (c) $N_{mon} = 3$.

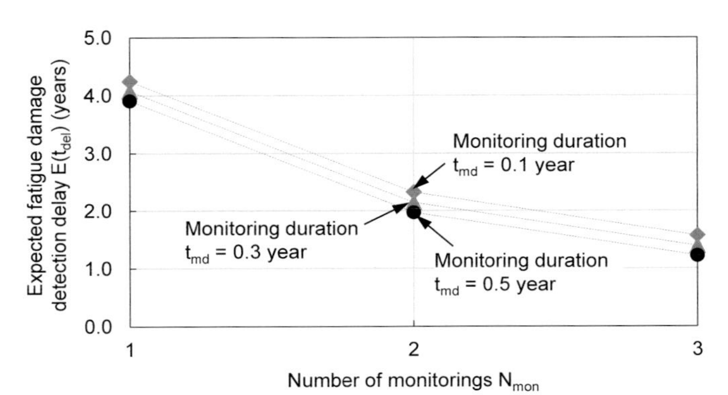

Figure 4.9 Relation between number of monitorings N_{mon} and expected fatigue crack damage detection delay $E(t_{del})$ for monitoring durations t_{md} = 0.1, 0.3 and 0.5 year.

given number of monitorings N_{mon} equal to 2 and monitoring duration t_{md} of 0.3 year, the first and second monitoring starting times (i.e., $t_{ms,1}$ and $t_{ms,2}$) are 4.24 years and 8.37 years, and the expected damage detection delay $E(t_{del})$ is 2.14 years, as indicated in Figure 4.8(b). If N_{mon} increases from 2 to 3, and t_{md} decreases from 0.3 to 0.1, $t_{ms,1}$, $t_{ms,2}$ and $t_{ms,3}$ are 3.48 years, 5.87 years and 9.89 years, respectively. The corresponding $E(t_{del})$ decreases from 2.14 years to 1.57 years (see Figure 4.8(c)). The relation among the number of monitorings N_{mon}, expected fatigue crack damage detection delay $E(t_{del})$, and monitoring durations t_{md} is illustrated in Figure 4.9. This figure shows that a reduction of the expected fatigue crack damage detection delay $E(t_{del})$ is caused by increasing the number of monitorings N_{mon} and/or increasing the monitoring duration t_{md}.

4.3.2 Minimizing Damage Detection Time-Based Probability of Failure

Minimizing the fatigue crack damage detection time-based probability of failure P_f (i.e., $O_{M,2}$) can be considered as the objective for optimum monitoring planning. For given number of monitorings N_{mon} and monitoring duration t_{md}, the corresponding single-objective optimization is formulated as

Find $\mathbf{t_{ms}} = \{t_{ms,1}, t_{ms,2}, \ldots, t_{ms,Nmon}\}$ (4.5a)

to minimize $P_f = P(t_{crt} - t_{det} < 0) = P(t_{mar} - t_{del} < 0)$ (4.5b)

The monitoring starting times $t_{ms,i}$ (years) are the design variables. The PDFs of t_{crt} and t_{mar} in Eq. (4.5b) are provided in Figure 3.7. The formulations of t_{det} and t_{del} for monitoring are described in Chapter 3 (see Eqs. 3.17 and 3.18). The constraints and given conditions used for this optimization problem are provided in Eqs. (4.4c) and (4.4d).

The monitoring plans based on $O_{M,2}$ are presented in Figure 4.10. When one-time monitoring for the monitoring duration $t_{md} = 0.1$ year is available, the monitoring should start at 6.21 years. If t_{md} increases from 0.1 year to 0.5 year, the required monitoring starting time becomes 5.93 years as shown in Figure 4.10(a). Furthermore, three-time monitoring with t_{md} = 0.5 year should be applied at 2.36 years, 4.13 years and 7.78 years, in order to obtain the minimum failure probability $P_f = 0.0255$. As shown in Figure 4.10, it can be seen that the monitoring starting times are delayed by reducing the monitoring duration t_{md}. P_f can be reduced by increasing the number of monitorings N_{mon} and/or increasing the monitoring duration t_{md}, as indicated in Figure 4.11.

4.4 Optimum Inspection and Monitoring Planning

The combined inspection and monitoring planning can be based on $O_{C,1}$ (i.e., minimizing the expected fatigue crack damage detection delay $E(t_{del})$) or $O_{C,2}$ (i.e., minimizing the fatigue crack damage detection time-based probability of failure P_f). When one inspection and one monitoring are available, the optimization problem for $O_{C,1}$ is formulated as

Find $t_{insp,1}$ and $t_{ms,1}$ (4.6a)

to minimize $E(t_{del})$ (4.6b)

such that $1 \text{ year} \leq t_{ms,i} - t_{insp,1} \leq 20$ years,

or $1 \text{ year} \leq t_{insp,1} - (t_{ms,i} + t_{md}) \leq 20$ years (4.6c)

given type of the inspection method and t_{md} (4.6d)

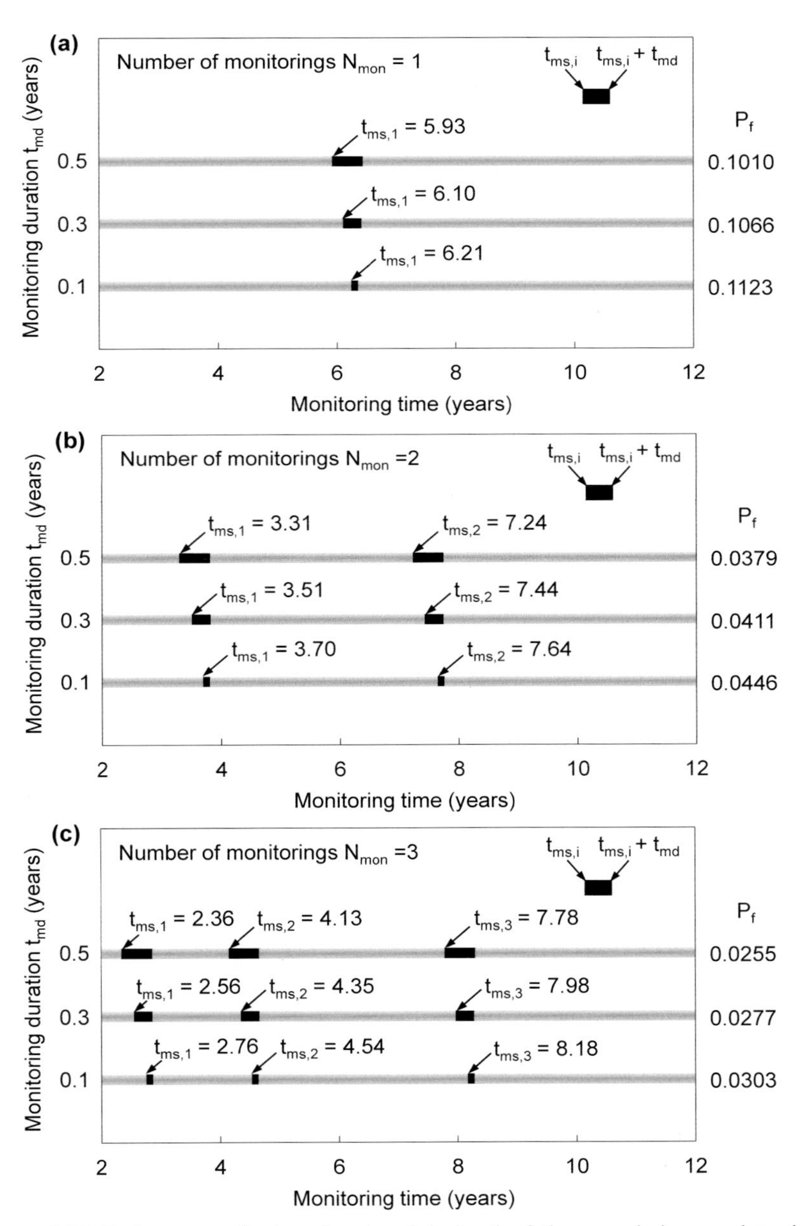

Figure 4.10 Optimum monitoring plans to minimize the fatigue crack damage detection time-based probability of failure P_f: (a) $N_{mon} = 1$; (b) $N_{mon} = 2$; (c) $N_{mon} = 3$.

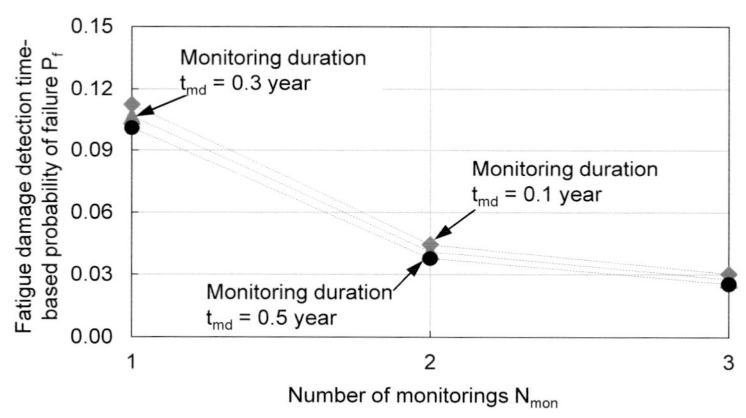

Figure 4.11 Relation between number of monitorings N_{mon} and damage detection time-based probability of failure P_f for monitoring durations t_{md} = 0.1, 0.3 and 0.5 year.

The design variables are the inspection and monitoring starting times as indicated in Eq. (4.6a). The formulation of $E(t_{del})$ for this optimization problem is provided in Eqs. (3.21) and (3.22). The time interval between inspection and monitoring has to be larger than 1 year and less than 20 years (see Eq. (4.6c)). The type of inspection method and monitoring duration t_{md} are given for the optimization problem.

The optimum combined inspection and monitoring plans are illustrated in Figure 4.12. If the monitoring with t_{md} = 0.5 year is applied after the EC inspection, the inspection and monitoring are required at 4.40 years and 8.17 years, respectively. The associated $E(t_{del})$ becomes 2.30 years, as shown in Figure 4.12(a). The reduction of t_{md} from 0.5 year to 0.1 year leads to an increase in $E(t_{del})$ from 2.30 years to 2.43 years. The minimum $E(t_{del})$ of 2.14 years can be obtained by applying the monitoring with t_{md} = 0.5 year at 4.19 years and the EC inspection at 9.05 years as shown in Figure 4.12(b). Figure 4.13 compares the expected damage detection delay $E(t_{del})$ for two cases: (a) inspection before monitoring and (b) inspection after monitoring. It is worth noting that case (b) leads to a reduction in $E(t_{del})$ compared to case (a).

The optimum inspection and monitoring planning based on $O_{C,2}$ (i.e., minimizing the fatigue crack damage detection time-based probability of failure P_f) is formulated as

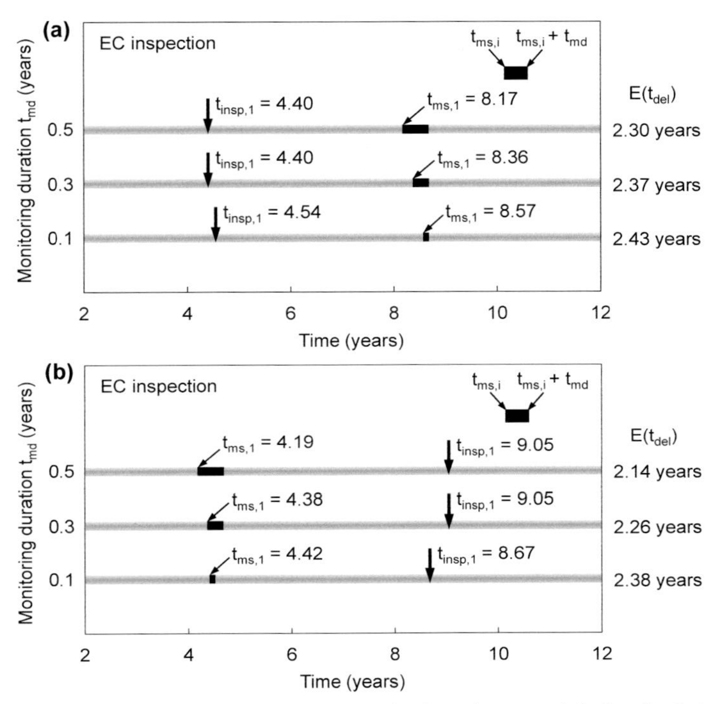

Figure 4.12 Optimum inspection and monitoring plans to minimize the fatigue crack damage detection delay $E(t_{del})$: (a) inspection → monitoring; (b) monitoring → inspection.

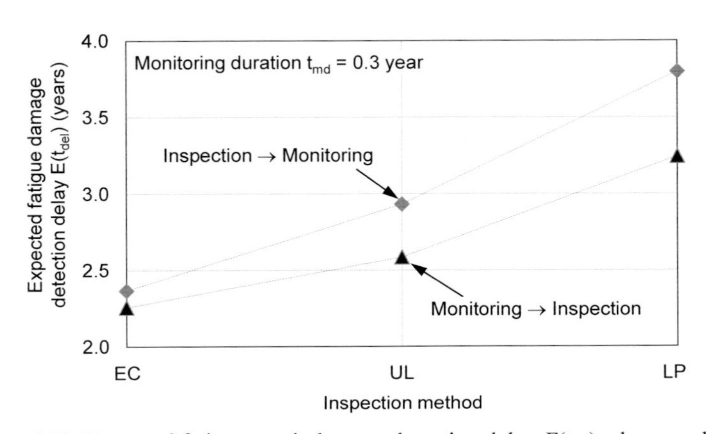

Figure 4.13 Expected fatigue crack damage detection delay $E(t_{del})$ when combined inspection/monitoring is used.

Find $t_{insp,1}$ and $t_{ms,1}$ (4.7a)

to minimize P_f (4.7b)

The constraints and given conditions are presented in Eqs. (4.6c) and (4.6d), respectively. The objective function P_f is based on Eqs. (3.21) and (3.22). The results of the optimization problem defined in Eq. (4.7) are illustrated in Figure 4.14. For given inspection method (i.e., UL inspection) and monitoring duration (i.e., $t_{md} = 0.5$ year), the minimum P_f of 0.0371 can be obtained by applying monitoring at 3.67 years and UL inspection at 8.53 years (see Figure 4.14(b)). Furthermore, Figure 4.15 shows that P_f for the monitoring before inspection is less than that for the inspection before monitoring.

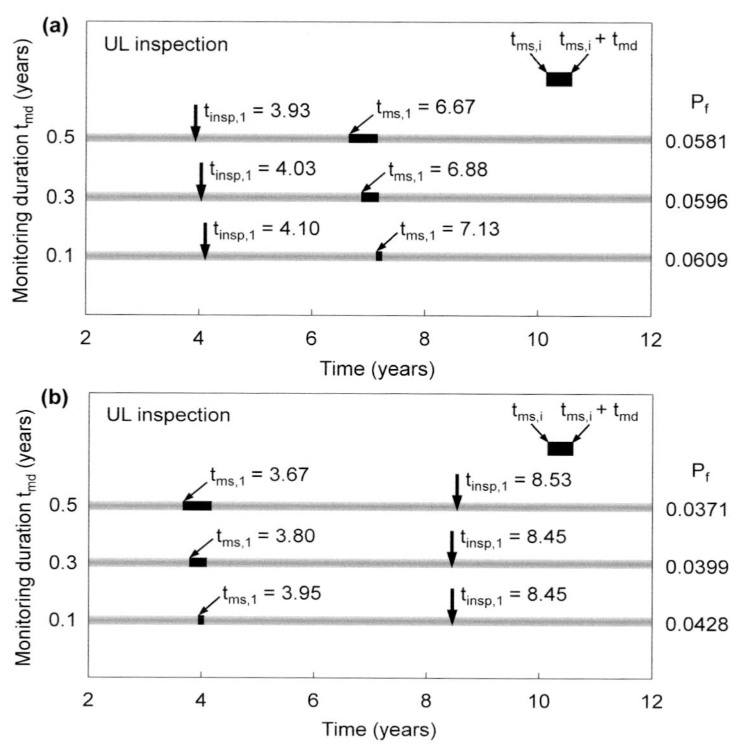

Figure 4.14 Optimum inspection and monitoring plans to the fatigue crack damage detection time-based probability of failure P_f: (a) inspection → monitoring; (b) monitoring → inspection.

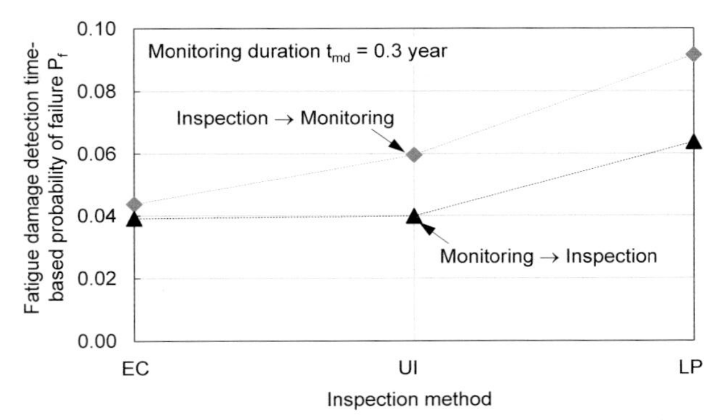

Figure 4.15 Fatigue crack damage detection time-based probability of failure P_f when combined inspection/monitoring is used.

4.5 Conclusions

This chapter addresses the single-objective optimum inspection and monitoring planning for service life management of fatigue-sensitive structures. The results associated with optimum inspection planning reveal that an increase in the number of inspections and/or improvement of inspection quality can lead to earlier first inspection time, increase of lifetime probability of fatigue crack damage detection, and reduction of both the expected fatigue crack damage detection delay and the fatigue crack damage detection time-based probability of failure. The monitoring planning with larger number of monitorings and longer monitoring duration also reduces both the expected fatigue crack damage detection delay and the fatigue crack damage detection time-based probability of failure. Since an increase in the number of inspections and/or improvement of inspection quality, and the increase in number of monitorings and/or monitoring duration can affect service life extension and life-cycle cost, the optimization process by considering the effect of inspection and monitoring on the service life extension and life-cycle cost needs to be further studied. Such an investigation will be presented subsequently.

Chapter **5**

Optimum Service Life and Life-Cycle Cost Management

CONTENTS

ABSTRACT

Chapter 5 provides the probabilistic single–objective optimum service life and life-cycle cost management for fatigue-sensitive civil and marine structures. The associated single-objective optimization problems are based on minimizing the expected maintenance delay, maximizing the expected extended service life, and minimizing the expected life-cycle cost. These objectives are formulated using the event tree, which considers the probability of fatigue damage detection and the effects of inspection and maintenance on the service life extension and life–cycle cost under uncertainty. The relations among the number of inspections and monitorings, expected maintenance delay, expected extended service life and expected life-cycle cost are investigated. By solving the single-objective optimization problem for a given number of inspections and inspection types, the optimum inspection times are obtained. For monitoring planning, optimum monitoring starting times are determined when the number of monitorings and monitoring durations are predefined. The approach presented is illustrated on an existing fatigue-sensitive bridge.

5.1 Introduction

Service life management to improve structural performance and safety and to extend the service life for fatigue-sensitive structures requires financial resources (Enright and Frangopol 1999a; Frangopol and Estes 1997; Frangopol and Maute 2003; Frangopol et al. 2004; Garbatov and Guedes Soares 2001). Efficient service life management should be based on the optimum allocation of financial resources (ASCE 2017; NCHRP 2003, 2012). Life-cycle cost analysis has been recognized as an essential tool for efficient service life management of deteriorating structures, and related approaches have been investigated and applied to various types of steel structures including bridges, offshore structures and naval structures (Frangopol 2011; Frangopol et al. 2017; Frangopol and Kim 2011, 2014b; Frangopol and Soliman 2016; Soliman et al. 2016). Efficient life-cycle cost management requires optimization to determine the optimum application of inspection and maintenance actions (Barone and Frangopol 2013; Barone et al. 2014; Bucher and Frangopol 2006; Dong and Frangopol 2015; Estes and Frangopol 1999, 2001; Miyamoto et al. 2000; Liu and Frangopol 2004; Lukic and Cremona 2001; Chung et al. 2006; Okasha and Frangopol 2009; Soliman et al. 2016). The optimization process can be based on single or multiple objectives considering the effect of inspection and maintenance on service life extension and life-cycle cost minimization (Kim and Frangopol 2017).

This chapter deals with the probabilistic single-objective optimum service life and life-cycle cost management for fatigue-sensitive structures. The associated objectives include minimizing the expected maintenance delay, maximizing the expected extended service life, and minimizing the expected life-cycle cost. The formulations of these objectives are based on the event tree considering the probability of fatigue damage detection and the effects of inspection and maintenance on service life extension and life-cycle cost under uncertainty. The relations between the number of inspections and monitorings, expected maintenance delay, expected extended service life and expected life-cycle cost are investigated. The single-objective optimization process is applied to establish the inspection and monitoring plans. As a result, optimum inspection times are obtained by solving the single-objective optimization problem for a given number of inspections and inspection type. For monitoring planning, optimum monitoring starting times are determined when the number of monitorings and monitoring durations are predefined. The approach presented in this chapter is illustrated on an existing fatigue-sensitive bridge.

5.2 Service Life and Life-Cycle Cost Under Uncertainty

Inspections are generally performed to detect and identify damage, and to provide information on structural performance deterioration. If the damage is detected and identified by an inspection, appropriate maintenance actions will be applied so that the structural performance deterioration can be delayed or structural performance be improved. As a result, the service life of a deteriorating structure can be extended (Thoft-Christensen and Sørensen 1987; Onoufriou and Frangopol 2002; Frangopol and Kim 2014b). The inspection and maintenance to extend the service life, affect the life-cycle cost. For this reason, the service life and life-cycle cost have to be taken into account in an integrated manner (Kim and Frangopol 2017, 2018a, 2018b).

5.2.1 *Effect of Inspection and Maintenance on Service Life and Life-Cycle Cost*

The effect of inspection and maintenance on the service life and cost using a decision tree is illustrated in Figure 5.1. For the purpose of illustration, the two inspections are applied at time $t_{insp,1}$ and $t_{insp,2}$, and the first inspection at $t_{insp,1}$ results in no maintenance. There are two reasons for no maintenance after the inspection: (a) damage is detected, but the degree of damage is too small to apply maintenance; and (b) no damage is detected. If there is no maintenance even though damage is detected, the degree of damage will be estimated by the second inspection at time $t_{insp,2}$, and a type of maintenance (i.e., no maintenance, and maintenances A and B) can be determined (see Branches 1, 2 and 3 in Figure 5.1). When the degree of damage is negligible, no maintenance is required. By applying maintenance A, deterioration of the structural performance is delayed during the time interval $t_{life,ex}$, and the service life is extended from the initial service life $t_{life,0}$ to $t_{life,0} + t_{life,ex}$. The required cost will be $C_{insp,1} + C_{insp,2} + C_{ma,A}$, where $C_{insp,i}$ is the cost for *i*th inspection, and $C_{ma,A}$ is the cost for maintenance A. Maintenance B leads to improving the structural performance from P_m to the initial structural performance P_0, and the service life and cost become

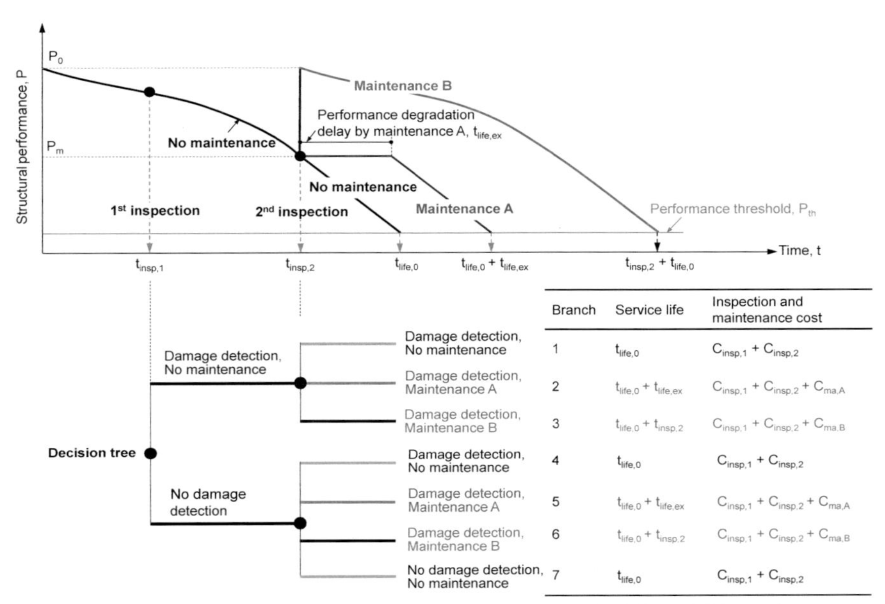

	Branch	Service life	Inspection and maintenance cost
Damage detection, No maintenance	1	$t_{life,0}$	$C_{insp,1} + C_{insp,2}$
Damage detection, Maintenance A	2	$t_{life,0} + t_{life,ex}$	$C_{insp,1} + C_{insp,2} + C_{ma,A}$
Damage detection, Maintenance B	3	$t_{life,0} + t_{insp,2}$	$C_{insp,1} + C_{insp,2} + C_{ma,B}$
Damage detection, No maintenance	4	$t_{life,0}$	$C_{insp,1} + C_{insp,2}$
Damage detection, Maintenance A	5	$t_{life,0} + t_{life,ex}$	$C_{insp,1} + C_{insp,2} + C_{ma,A}$
Damage detection, Maintenance B	6	$t_{life,0} + t_{insp,2}$	$C_{insp,1} + C_{insp,2} + C_{ma,B}$
No damage detection, No maintenance	7	$t_{life,0}$	$C_{insp,1} + C_{insp,2}$

Figure 5.1 Effect of inspection and maintenance on service life and life-cycle cost.

Color version at the end of the book

$t_{insp,2} + t_{life,0}$ and $C_{insp,1} + C_{insp,2} + C_{ma,B}$, respectively. When the damage is detected by the second inspection at $t_{insp,2}$ without damage detection at the first inspection time $t_{insp,1}$, maintenances A and B, and no maintenance can be applied according to the degree of damage. The associated events are considered in Branches 4, 5 and 6. If the damage is not detected at the second inspection time $t_{insp,2}$, there will be no maintenance and no service life extension, and the service life and cost will be $t_{life,0}$ and $C_{insp,1} + C_{insp,2}$, respectively (see Branch 7).

5.2.2 *Maintenance Delay, Service Life and Life-Cycle Cost for Inspection*

The maintenance delay, extended service life and life-cycle cost for fatigue-sensitive structures can be formulated using the decision tree illustrated in

Figure 5.2, which is generalized considering the effect of inspection and maintenance on service life and life-cycle cost presented in Figure 5.1. Figure 5.2 addresses all the possible events when a single scheduled inspection is performed to detect the damage occurring at time t_{oc}. The maintenance delay t_{mdl}, service life t_{life} and life-cycle cost C_{lcc} of each branch in Figure 5.2 are provided in Table 5.1. The branch BR_1 is associated with the event that the inspection time $t_{insp,1}$ is applied after the initial service life $t_{life,0}$ (i.e., $t_{insp,1} \geq t_{life,0}$), and there will be no service life extension. The corresponding t_{mdl}, t_{life} and C_{lcc} will be $t_{life,0} - t_{oc}$, $t_{life,0}$ and $C_{insp,1} + C_{fail,0}$,

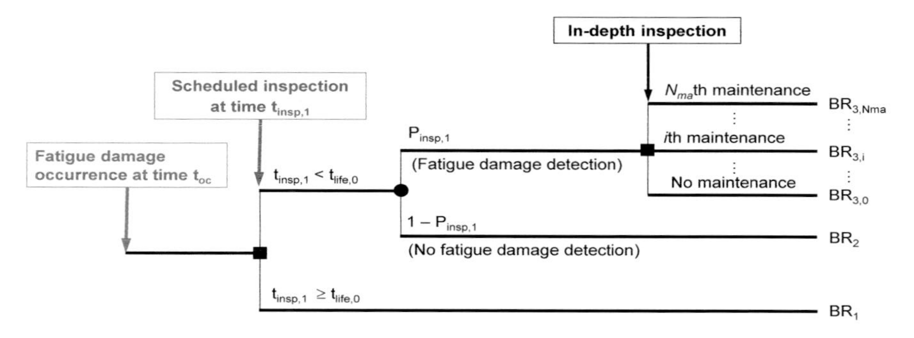

Figure 5.2 Decision tree for formulation of maintenance delay, service life extension and life-cycle cost.

Table 5.1 Maintenance delay, extended service life and life-cycle of each branch in Figure 5.2.

Branch	Maintenance delay t_{mdl}	Service life t_{life}	Life-cycle cost C_{lcc}
BR_1	$t_{life,0} - t_{oc}$	$t_{life,0}$	$C_{insp,1} + C_{fail,0}$
BR_2	$t_{life,0} - t_{oc}$	$t_{life,0}$	$C_{insp,1} + C_{fail,o}$
$BR_{3,0}$	$t_{life,0} - t_{oc}$	$t_{life,0}$	$C_{insp,1} + C_{insp,d} + C_{fail,0}$
$BR_{3,i}$	$t_{insp,1} - t_{oc}$	$t_{life,i}$	$C_{insp,1} + C_{insp,d} + C_{ma,i} + C_{fail,i}$
$BR_{3,Nma}$	$t_{insp,1} - t_{oc}$	$t_{life,Nma}$	$C_{insp,1} + C_{insp,d} + C_{ma,Nma} + C_{fail,Nma}$

respectively, where $C_{fail,0}$ is the expected failure cost without maintenance. If the inspection before the initial service life $t_{life,0}$ detects no damage, there will be no maintenance and no service life extension (see the branch BR_2 in Figure 5.2 and Table 5.1). If the fatigue crack damage is detected before the end of service life, an in-depth inspection will be performed to investigate the degree of damage (i.e., fatigue crack size), and determine the type of maintenance. The ith maintenance type will cause the maintenance delay $t_{insp,i} - t_{oc}$, extended service life $t_{life,i}$ and life-cycle cost $C_{insp,1} + C_{insp,d} + C_{ma,i} + C_{fail,i}$, where $C_{insp,d}$ = in-depth inspection cost, $C_{fail,i}$ = expected failure cost by the ith maintenance application. This corresponds to the branch $BR_{3,i}$ in Figure 5.2 and Table 5.1.

Considering all the branches of the decision tree in Figure 5.2, the maintenance delay t_{mdl} for a single scheduled inspection at time $t_{insp,1}$ is determined as (Kim et al. 2013)

$$t_{mdl} = t_{life,0} - t_{oc} \qquad \text{for } t_{insp,1} \geq t_{life,0} \tag{5.1a}$$

$$t_{mdl} = P_{insp,1} \cdot t_{mdl,1} + (1 - P_{insp,1}) \cdot (t_{life,0} - t_{oc}) \qquad \text{for } t_{insp,1} < t_{life,0} \tag{5.1b}$$

where $P_{insp,1}$ = probability of fatigue crack damage detection; and $t_{mdl,1}$ = maintenance delay after fatigue crack damage detection by the inspection at time $t_{insp,1}$, which is expressed as

$$t_{mdl,1} = t_{life,0} - t_{oc} \qquad \text{for } a \leq a_{ma,0} \tag{5.2a}$$

$$t_{mdl,1} = t_{insp,1} - t_{oc} \qquad \text{for } a_{ma,0} < a \tag{5.2b}$$

where $a_{ma,0}$ = critical fatigue crack size requiring a maintenance. When the crack size is less than $a_{ma,0}$, there will be no maintenance, and the associated delay becomes $t_{life,0} - t_{oc}$ (see the branch $BR_{3,0}$ in Figure 5.2 and Table 5.1).

Furthermore, the extended service life t_{life} for a single scheduled inspection can be expressed as (Kim et al. 2011, 2013)

$$t_{life} = t_{life,0} \qquad \text{for } t_{insp,1} \geq t_{life,0} \tag{5.3a}$$

$$t_{life} = P_{insp,1} \cdot (t_{life,0} + t_{exd}) + (1 - P_{insp,1}) \cdot t_{life,0} \qquad \text{for } t_{insp,1} < t_{life,0} \tag{5.3b}$$

where t_{exd} = service life extension after the fatigue crack damage detection by the inspection at time $t_{insp,1}$. When N_{ma} types of maintenance are available, the service life extension t_{exd} in Eq. (5.3b) can be expressed as

$$t_{exd} = 0 \qquad \text{for } a \leq a_{ma,0} \tag{5.4a}$$

$$t_{exd} = t_{exd,i} \qquad \text{for } a_{ma,i-1} < a \leq a_{ma,i} \tag{5.4b}$$

where $a_{ma,i}$ = critical fatigue crack size for ith maintenance type. $t_{exd,i}$ = service life extension by applying the ith maintenance type among N_{ma} types. The ith maintenance type is used when the crack size a is between $a_{ma,i-1}$ and $a_{ma,i}$, as indicated in Eq. (5.4b).

Also, based on the decision tree in Figure 5.2, the life-cycle cost C_{lcc} for a single scheduled inspection is formulated as (Thoft-Christensen and Sørensen 1987; Frangopol et al. 1997)

$$C_{lcc} = C_{insp,1} + C_{fail,0} \qquad \text{for } t_{insp,1} \geq t_{life,0} \tag{5.5a}$$

$$C_{lcc} = P_{insp,1} \cdot (C_{insp,1} + C_{insp,d} + C_{dmg}) + (1 - P_{insp,1}) \cdot (C_{insp,1} + C_{fail,0}) \qquad \text{for } t_{insp,1} < t_{life,0} \tag{5.5b}$$

where $C_{fail,0}$ = expected failure cost without maintenance; and C_{dmg} = cost after damage detection expressed as

$$C_{dmg} = C_{fail,0} \qquad \text{for } a \leq a_{ma,0} \tag{5.6a}$$

$$C_{dmg} = C_{ma,i} + C_{fail,i} \qquad \text{for } a_{ma,i-1} < a \leq a_{ma,i} \tag{5.6b}$$

If the detected crack size is less than $a_{ma,0}$, there will be no maintenance, and C_{dmg} will be equal to $C_{fail,0}$. The detected crack size a between $a_{ma,i-1}$ and $a_{ma,i}$ leads to the ith maintenance, and as a result, the expected failure cost becomes $C_{fail,i}$. The expected failure cost for the ith maintenance action $C_{fail,i}$ is estimated as

$$C_{fail,i} = C_{loss} \cdot P(t_{life} \leq t^*) \tag{5.7}$$

where C_{loss} is the expected monetary loss due to the structural failure, t_{life} is the extended service life defined in Eq. (5.3), and t^* is the predefined target service life. $C_{fail,0}$ in Eqs. (5.5a) and (5.6a) is computed as $C_{loss} \cdot P(t_{life,0} \leq t^*)$. In this manner, the maintenance delay, extended service life, and life-cycle for multiple inspections can be formulated. When the uncertainties associated with the fatigue crack occurrence and propagation, damage detection by an inspection method, and maintenance effect on the service life are considered, the expected maintenance delay $E(t_{mdl})$, expected extended service life $E(t_{life})$ and expected life-cycle cost $E(C_{lcc})$ can be obtained.

5.2.3 *Maintenance Delay, Service Life and Life-Cycle Cost for Monitoring*

When monitoring is applied for service life management of fatigue-sensitive structures, the maintenance delay, extended service life and life-cycle cost formulated in Eqs. (5.1)–(5.7) can be modified by assuming that the probability of damage detection during monitoring duration is perfect (Kim and Frangopol 2018a). The maintenance delay t_{mdl} for a single monitoring is expressed based on Eqs. (5.1) and (5.2) as:

$$t_{mdl} = t_{life,0} - t_{oc} \quad \text{for } t_{ms,1} \geq t_{life,0} \tag{5.8a}$$

$$t_{mdl} = t_{mdl,1} \quad \text{for } t_{ms,1} < t_{life,0} \tag{5.8b}$$

where $t_{ms,1}$ = first monitoring starting time; and $t_{mdl,1}$ = maintenance delay after fatigue crack damage detection by monitoring. Considering the critical crack size $a_{ma,0}$ requiring maintenance actions, $t_{mdl,1}$ is

$$t_{mdl,1} = t_{life,0} - t_{oc} \quad \text{for } a \leq a_{ma,0} \tag{5.9a}$$

$$t_{mdl,1} = t_{ms,1} - t_{oc} \quad \text{for } a_{ma,0} < a \tag{5.9b}$$

Also, the extended service life t_{life} for a single monitoring is formulated as

$$t_{life} = t_{life,0} \quad \text{for } t_{ms,1} \geq t_{life,0} \tag{5.10a}$$

$$t_{life} = t_{life,0} + t_{exd} \qquad \text{for } t_{ms,1} < t_{life,0} \tag{5.10b}$$

The service life extension after the damage detection t_{exd} in Eq. (5.10b) is estimated using Eq. (5.4). Similarly to the formulation of Eq. (5.5), the life-cycle cost C_{lcc} for a single monitoring is

$$C_{lcc} = C_{mon} + C_{fail,0} \qquad \text{for } t_{ms,1} \geq t_{life,0} \tag{5.11a}$$

$$C_{lcc} = C_{mon} + C_{insp,d} + C_{dmg} \qquad \text{for } t_{ms,1} < t_{life,0} \tag{5.11b}$$

The monitoring cost C_{mon} in Eq. (5.11b) is estimated as

$$C_{mon} = C_{mon,i} + C_{mon,o} \cdot t_{md} \tag{5.12}$$

where $C_{mon,i}$ = initial monitoring cost for installation of the monitoring system; $C_{mon,o}$ = monitoring cost for operation and maintenance, which may be proportional to monitoring duration t_{md}. In order to compute the cost after damage detection C_{dmg} in Eq. (5.11b), Eq. (5.6) is used.

5.2.4 Application to a Fatigue-Sensitive Bridge

In this chapter, a fatigue-sensitive detail of the I-64 Bridge over the Kanawha River in West Virginia is investigated as an illustrative example. This fatigue-sensitive detail is the bottom web gap between the bottom flange of the exterior girder and the transverse connecting plate of the exterior girder, as shown in Figure 5.3. In order to predict the fatigue crack propagation over time, Eq. (3.4) is applied with the probabilistic and deterministic variables defined in Table 5.2. The geometry function $Y(a)$ in Eq. (3.4) is estimated based on a semi-elliptical edge crack (Fisher 1984). The critical cracks resulting in fatigue damage occurrence and fatigue failure are assumed to be 1 *mm* and 10.16 *mm* (i.e., thickness of the web), respectively (Kim and Frangopol 2018b). In this illustrative example, the ultrasonic (UL) inspection is applied to detect the fatigue crack damage. The single type of maintenance (i.e., cutting out and re-fabricating parts of elements) is applied to return to the initial condition, when the crack size is reaches $a_{ma,0}$. The probability of fatigue damage detection is based on

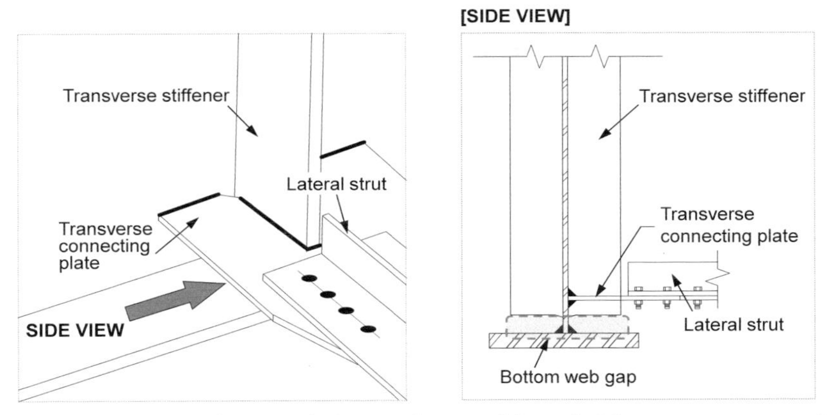

Figure 5.3 Bottom gap between the bottom flange and the end of the transverse connecting plate of a steel bridge (adapted from Connor and Fisher 2011; Soliman et al. 2013a).

Table 5.2 Variables for crack growth model at the gap between the bottom flange and the end of the transverse connecting plate of a steel bridge (see Figure 5.3).

Variables	**Notation**	**Units**	**Mean**	***COV**	**Type of distribution**
Initial crack size	a_o	mm	0.5	0.2	Lognormal
Annual number of cycles	N_{an}	cycles/ year	2.74×10^6	0.1	Lognormal
Stress range	S_r	MPa	34.5	0.1	Weibull
Material crack growth parameter	C	-	2.18×10^{-13}	0.2	Lognormal
Material exponent	m	-	3	-	Deterministic

*COV: coefficient of variation.
Based on information in Connor and Fisher (2001) and Soliman et al. (2013a).

the cumulative lognormal distribution function form with the parameters provided in Table 4.1. The required costs to formulate the life-cycle cost C_{lcc} for inspection and monitoring are presented in Table 5.3. Herein, the UL inspection times for N_{insp} = 1, 2 and 3 are {$t_{insp,1}$ = 5 years}, {$t_{insp,1}$ = 5 years; $t_{insp,2}$ = 10 years} and {$t_{insp,1}$ = 5 years; $t_{insp,2}$ = 10 years; $t_{insp,3}$ = 15 years}, respectively. The monitoring starting times for N_{mon} = 1, 2 and 3

Table 5.3 Costs for life-cycle cost estimation.

	In-depth inspection cost $C_{insp,d}$	Maintenance cost $C_{ma,i}$	Expected monetary loss C_{loss}	Scheduled UL inspection cost $C_{insp,i}$	Initial monitoring cost $C_{mon,i}$	Monitoring cost for operation and maintenance $C_{mon,o}$
Cost (USD)	15,000	50,000	1,000,000	8,500	15,000	1,000/week

Based on information in Kim and Frangopol (2018a) and (2018b).

are $\{t_{ms,1} = 5$ years$\}$, $\{t_{ms,1} = 5$ years; $t_{ms,2} = 10$ years$\}$ and $\{t_{ms,1} = 5$ years; $t_{ms,2} = 10$ years; $t_{ms,3} = 15$ years$\}$, respectively, and the monitoring duration t_{md} of 0.5 year is applied to this illustrative example.

Figure 5.4 shows the effects of number of inspections N_{insp} and crack size requiring a maintenance $a_{ma,0}$ on the expected maintenance delay $E(t_{mdl})$, expected extended service life $E(t_{life})$, and expected life-cycle cost $E(C_{lcc})$. For the crack size requiring a maintenance $a_{ma,0} = 1.5$ mm, it can be observed that an increase in N_{insp} leads to a decrease of $E(t_{mdl})$, increase of $E(t_{life})$ and decrease of $E(C_{lcc})$. However, when $a_{ma} = 0$ mm is applied, $E(C_{lcc})$ associated with $N_{insp} = 2$ is the smallest among $N_{insp} = 1$, 2 and 3 as shown in Figure 5.4(c). Furthermore, the change of $a_{ma,0}$ from 0.0 mm to 1.5 mm results in an increase of $E(t_{mdl})$, decrease of $E(t_{life})$ and increase of $E(C_{lcc})$.

The expected maintenance delay $E(t_{mdl})$ and expected extended service life $E(t_{life})$ for inspection and monitoring are compared in Figure 5.5. When the monitoring with $t_{md} = 0.5$ is used instead of the UL inspection, less $E(t_{mdl})$ and larger $E(t_{life})$ can be obtained. In Figure 5.6, the effect of N_{insp} and N_{mon} on $E(C_{lcc})$ are compared with three difference expected monetary losses (i.e., $C_{loss} = \$10^5$, $\$10^6$ and $\$10^7$). For $C_{loss} = \$10^5$, the monitoring with $t_{md} = 0.5$ years results in larger $E(C_{lcc})$ than UL inspection (see Figure 5.6(a)). However, if C_{loss} is equal to $\$10^7$, $E(C_{lcc})$ corresponding to the monitoring with $t_{md} = 0.5$ years is less than one for the UL inspection (see Figure 5.6(c)). Moreover, while an increase in the number of inspections or monitorings for $C_{loss} = \$10^5$ leads to larger $E(C_{lcc})$, a reduced $E(C_{lcc})$ can be obtained by increasing the number of inspections or monitorings for $C_{loss} = \$10^6$ and $\$10^7$.

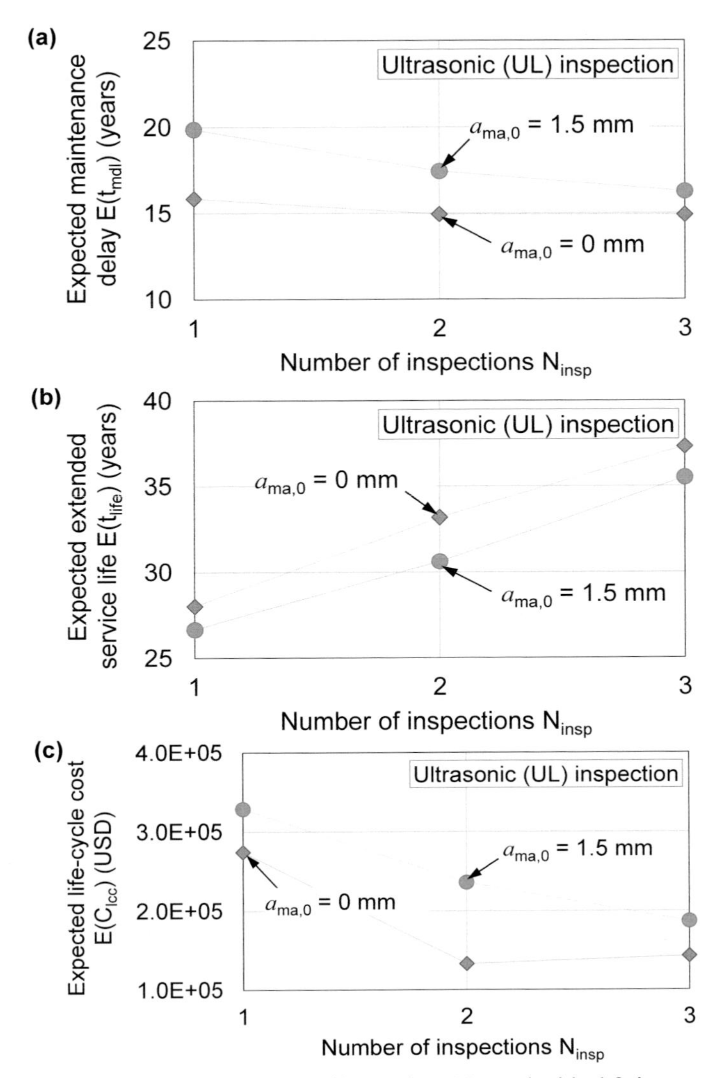

Figure 5.4 Relation among number of inspections N_{insp} and critical fatigue crack size requiring a maintenance action $a_{ma,0}$: (a) expected maintenance delay $E(t_{mdl})$; (b) expected extended service life $E(t_{life})$; (c) expected life-cycle cost $E(C_{lcc})$.

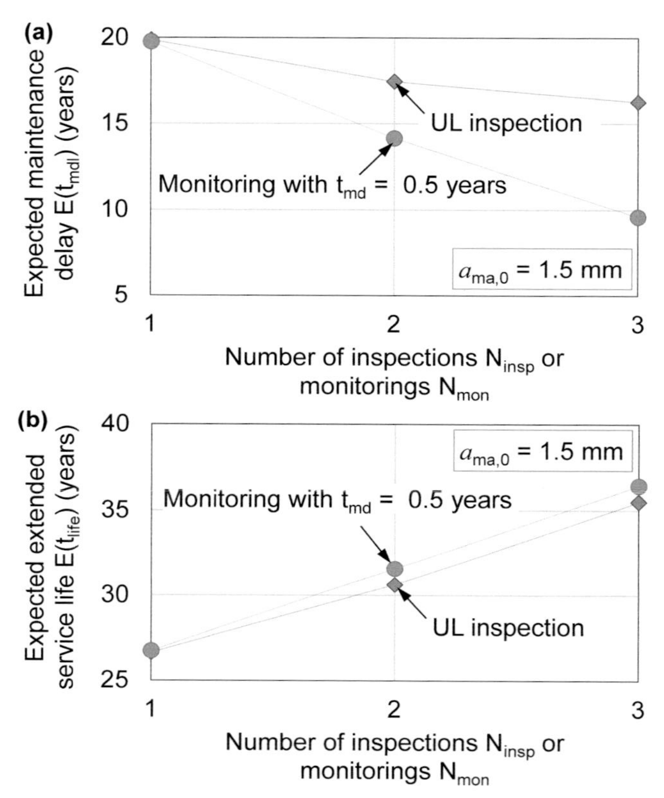

Figure 5.5 Effect of number of inspections and monitoring on: (a) expected maintenance delay $E(t_{mdl})$; (b) expected extended service life $E(t_{life})$.

5.3 Probabilistic Optimum Inspection and Monitoring Planning

The single-objective optimization for inspection and monitoring planning can be based on the three individual objectives of minimizing the expected maintenance delay $E(t_{mdl})$, maximizing the expected extended service life $E(t_{life})$ and minimizing the expected life-cycle cost $E(C_{lcc})$. The formulations of these objective functions $E(t_{mdl})$, $E(t_{life})$ and $E(C_{lcc})$ are addressed in the previous section. The solution of the single-objective optimization

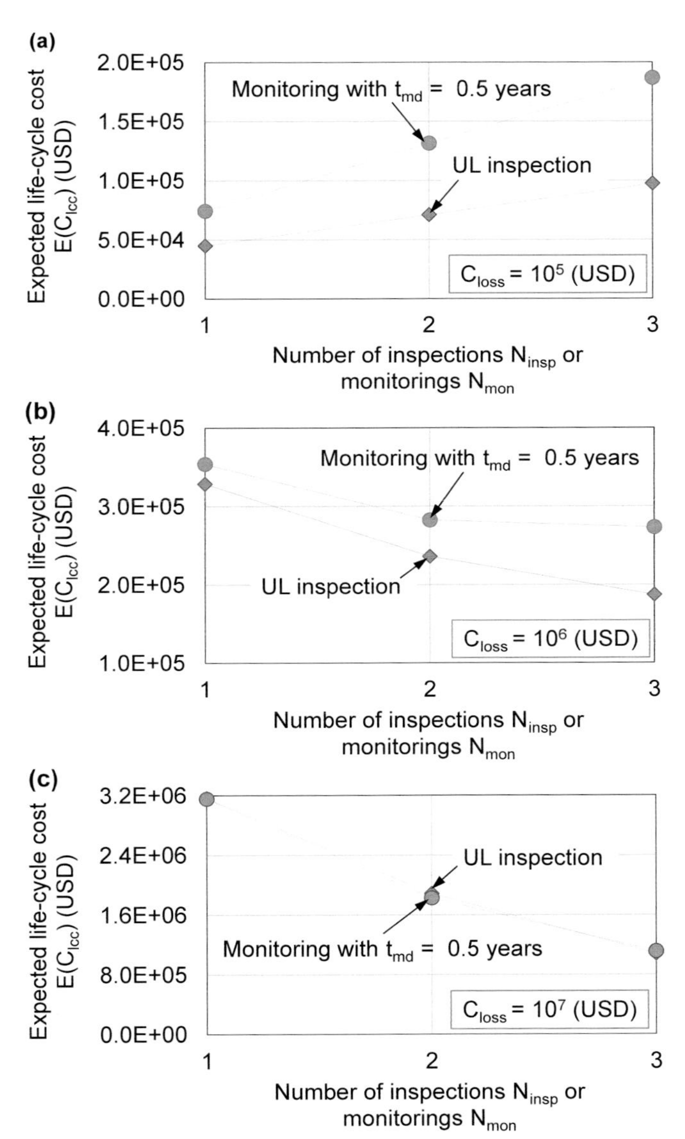

Figure 5.6 Effect of number of inspections and monitoring on expected life-cycle cost $E(C_{lcc})$: (a) $C_{loss} = 10^5$ (USD); (b) $C_{loss} = 10^6$ (USD); (c) $C_{loss} = 10^7$ (USD).

problem for inspection planning consists of the inspection times (i.e., $t_{insp,1}$, ..., $t_{insp,Ninsp}$). The optimum monitoring starting time (i.e., $t_{ms,1}$, ..., $t_{ms,Nmon}$) is determined by solving the single-objective optimization problem for monitoring planning. The single-objective optimization problems in this chapter are solved using the constrained nonlinear minimization algorithm provided in MATLAB® version 2016b (MathWorks 2016). In order to check whether the optimum solution is global, the genetic algorithm (GA) in MATLAB version 2016b is used. The single-objective optimization for inspection and monitoring planning is applied to a fatigue-sensitive bridge presented previously in Sub-section 5.2.4. It should be noted that the costs to estimate the expected life-cycle cost $E(C_{lcc})$ are provided in Table 5.3.

5.3.1 *Single-Objective Probabilistic Optimization for Inspection Planning*

The optimum inspection planning based on the single objective is formulated as

$$\text{Find } \mathbf{t}_{\mathbf{insp}} = \{t_{insp,1}, t_{insp,2}, \ldots, t_{insp,Ninsp}\} \tag{5.13a}$$

$$\text{for } \mathrm{O}_{\mathrm{I},4}, \mathrm{O}_{\mathrm{I},5}, \text{ or } \mathrm{O}_{\mathrm{I},6} \tag{5.13b}$$

$$\text{such that } t_{insp,i} - t_{insp,i-1} \geq 1 \text{ year and } t_{insp,i} \leq 20 \text{ years} \tag{5.13c}$$

$$\text{given } N_{insp} \text{ and type of the inspection method} \tag{5.13d}$$

where $\mathbf{t}_{\mathbf{insp}}$ = vector of design variables (i.e., inspection times (years)); $\mathrm{O}_{\mathrm{I},4}$ = minimizing the expected maintenance delay $E(t_{mdl})$; $\mathrm{O}_{\mathrm{I},5}$ = maximizing the expected extended service life $E(t_{life})$; $\mathrm{O}_{\mathrm{I},6}$ = minimizing the expected life-cycle cost $E(C_{lcc})$ for inspection; $t_{insp,i}$ = ith inspection time (years). The time interval between inspections has to be larger than 1 year, and the inspection time should be less than 20 years (see Eq. (5.13c)). The number of inspections N_{insp} and type of the inspection method are indicated in Eq. (5.13d).

The optimum UL inspection plans obtained from the optimization problem defined in Eq. (5.13) are illustrated in Figure 5.7. The inspection at $t_{insp,1}$ =

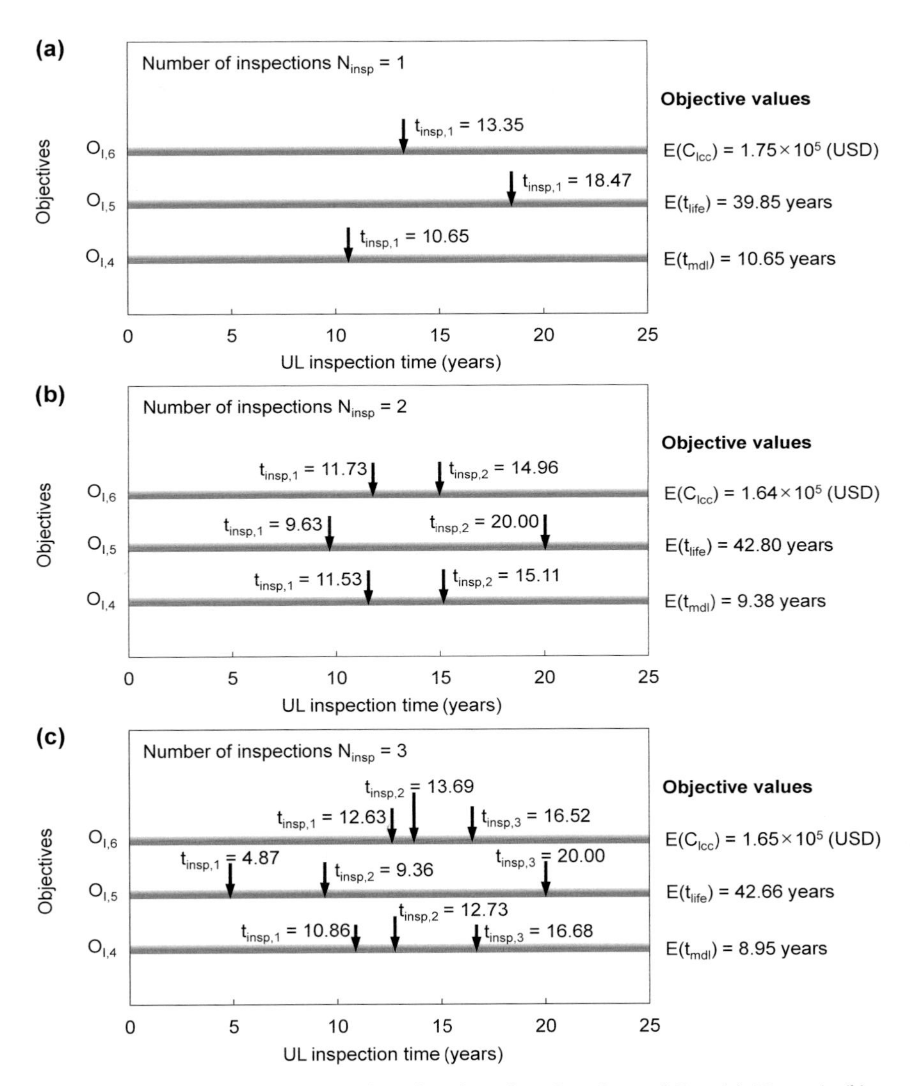

Figure 5.7 Optimum UL inspection plans based on $O_{I,4}$, $O_{I,5}$ and $O_{I,6}$: (a) $N_{insp} = 1$; (b) $N_{insp} = 2$; (c) $N_{insp} = 3$.

10.65 years results in the minimum $E(t_{mdl})$ of 10.65 years. The minimum $E(C_{lcc}) = \$1.75 \times 10^5$ can be estimated by applying the UL inspection at $t_{insp,1}$ = 13.35 years as indicated in Figure 5.7(a). If the number of inspections increases from one to two, the minimum $E(t_{mdl})$ and $E(C_{lcc})$ can be reduced by 11.9% (i.e., from 10.65 years to 9.38 years) and 6.3% (i.e., from $\$1.75 \times 10^5$ to $\$1.64 \times 10^5$). As shown in Figure 5.7(c), the three inspections at $t_{insp,1} = 4.87$ years, $t_{insp,2} = 9.36$ years and $t_{insp,3} = 20.00$ years can lead to the maximum expected extended service life $E(t_{life})$ of 42.66 years.

5.3.2 *Single-Objective Probabilistic Optimization for Monitoring Planning*

The formulation of the single-objective optimization for monitoring planning is given as

$$\text{Find } \mathbf{t_{ms}} = \{t_{ms,1}, t_{ms,2}, \ldots, t_{ms,Nmon}\} \quad (5.14a)$$

$$\text{for } O_{M,4}, O_{M,5}, \text{ or } O_{M,6} \quad (5.14b)$$

$$\text{such that } 1 \text{ year} \leq t_{ms,i} - (t_{ms,i-1} + t_{md}) \leq 20 \text{ years} \quad (5.14c)$$

$$\text{given } N_{mon} \text{ and } t_{md} \quad (5.14d)$$

where $\mathbf{t_{ms}}$ = set of design variables (i.e., monitoring starting times (years)); $O_{M,4}$ = minimizing $E(t_{mdl})$; $O_{M,5}$ = maximizing $E(t_{life})$; and $O_{M,6}$ = minimizing $E(C_{lcc})$ for monitoring. As indicated in Eq. (5.14c), the non-monitoring time interval (i.e., $t_{ms,i} - (t_{ms,i-1} + t_{md})$) has to be larger than 1 year and less than 20 years. The number of monitorings N_{mon} and monitoring duration t_{md} are given for this optimization problem in Eq. (5.14d).

Figure 5.8 illustrates the single-objective optimum monitoring plans with $t_{md} = 0.3$ year by solving the optimization problem formulated in Eq. (5.14). The optimum monitoring plans with the monitoring times $N_{mon} = 1$, 2 and 3 are shown in Figures 5.8(a), 5.8(b) and 5.8(c), respectively. In order to minimize the expected maintenance delay $E(t_{mdl})$ for $N_{mon} = 2$ and $t_{md} = 0.3$ year, the monitorings have to be performed at $t_{ms,1} = 11.51$ years and $t_{ms,2} = 17.56$ years, and the corresponding $E(t_{mdl})$ becomes 9.51 years. If the three monitorings (i.e., $N_{mon} = 3$) are available, the monitoring starting times should be 10.06 years, 13.91 years, and 19.60 years to minimize $E(t_{mdl})$. By comparing the objective values $E(t_{mdl})$, $E(t_{life})$ and $E(C_{lcc})$ in Figures 5.8(a), 5.8(b) and 5.8(c), it can be seen that an increase in the number of

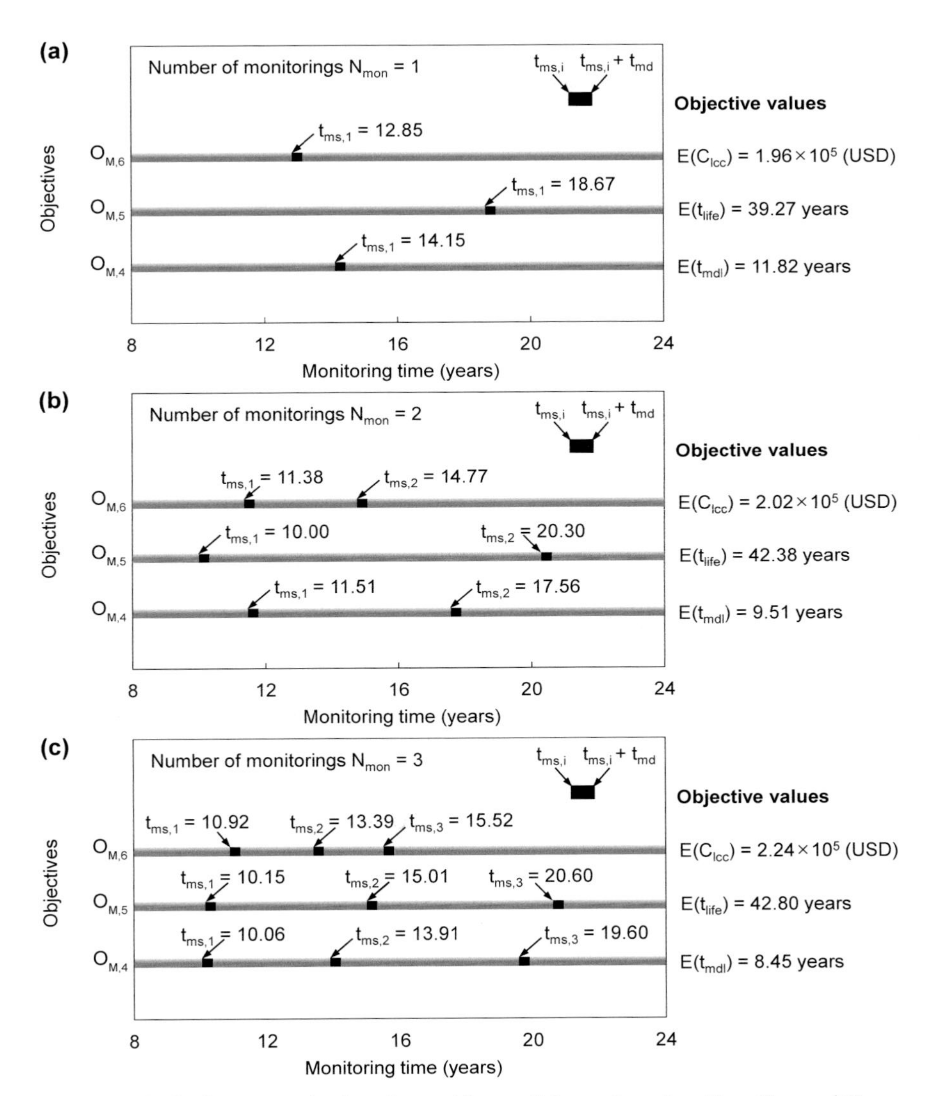

Figure 5.8 Optimum monitoring plans with $t_{md} = 0.3$ year based on $O_{M,4}$, $O_{M,5}$ and $O_{M,6}$: (a) $N_{mon} = 1$; (b) $N_{mon} = 2$; (c) $N_{mon} = 3$.

monitorings can lead to reduction of the minimum $E(t_{mdl})$, increase of the maximum $E(t_{life})$ and increase of the minimum $E(C_{lcc})$.

5.4 Conclusions

In this chapter, the probabilistic single-objective optimum service life and life-cycle cost management is addressed. The effects of inspection and maintenance on service life extension and life-cycle cost are implemented in a decision tree model to formulate the probabilistic objective functions such as expected maintenance delay, expected extended service life and expected life-cycle cost. From the results in this chapter, it can be seen that (a) the number of inspections, (b) inspection types, (c) number of monitorings, (d) monitoring duration, (e) critical fatigue crack size requiring a maintenance action, and (f) monetary loss due to failure affect the expected maintenance delay, expected extended service life and expected life-cycle cost. The optimum inspection and monitoring planning depends on the objective function. For example, the optimum inspection plan based on minimizing the expected maintenance delay is not identical to the optimum inspection plan based on maximizing the expected extended service life or minimizing the expected life-cycle cost. For this reason, it is necessary to integrate the multiple objectives for more rational inspection and monitoring planning. Such an investigation will be addressed subsequently.

Chapter **6**

Multi-Objective Probabilistic Life-Cycle Optimization

CONTENTS

ABSTRACT

When multiple objectives for optimum life–cycle management strategies are available, a successful integration of these objectives can be achieved, and multiple trade–off solutions can be obtained using a multi-objective optimization process. In Chapter 6, the multi-objective probabilistic optimum inspection and monitoring planning for fatigue-sensitive structures is investigated. The objective functions used in this chapter are the lifetime probability of fatigue crack damage detection, expected damage detection delay, damage detection time–based probability of failure, expected maintenance delay, expected extended service life and expected life-cycle cost. The bi- tri- and quad-objective and more than four objectives optimization problems are investigated. Representative solutions from the Pareto optimal set associated with the multi-objective optimization processes are compared. The multi-objective optimum inspection and monitoring planning approach presented, is applied to the fatigue-sensitive details of a steel bridge and a ship hull structure.

6.1 Introduction

Recently, significant efforts have been made to develop approaches for optimum life-cycle management strategies of fatigue-sensitive structures. Various demands on structural performance, safety, risk and financial resources have led to the development of new concepts and approaches for optimum life-cycle management strategies (Flintsch and Chen 2004; Kabir et al. 2014; Kim and Frangopol 2018a, 2018b). Several objectives considering both cost and structural performance were introduced

(Frangopol 2011; Frangopol and Soliman 2016). When multiple objectives for optimum life-cycle management strategies are available, a successful integration of these objectives can be achieved (Liu and Frangopol 2005b, 2006; Frangopol and Liu 2007; Orcesi and Frangopol 2011b; Sabatino et al. 2016). The multiple trade-off solutions can be obtained through a multi-objective optimization process (Deb 2001). The decision makers can select a rational solution by using additional information on life-cycle management (Fonseca and Fleming 1998; Yeh 2002).

In this chapter, the multi-objective probabilistic optimum inspection and monitoring planning for fatigue-sensitive structures is investigated. The objective functions used in this chapter are the lifetime probability of fatigue crack damage detection P_{det}, expected damage detection delay $E(t_{del})$, damage detection time-based probability of failure P_f, expected maintenance delay $E(t_{mdl})$, expected extended service life $E(t_{life})$ and expected life-cycle cost $E(C_{lcc})$. As indicated in Chapters 4 and 5, a single-objective optimization can provide its own inspection and/or monitoring plans. This chapter shows that multiple objective functions can be integrated through a multi-objective optimization process. The bi-, tri- and quad-objective and optimization problems with more than four objectives are investigated for inspection and monitoring planning. Several representative solutions from the Pareto optimal set associated with the multi-objective optimization processes are compared. In this chapter, a multi-objective optimum inspection planning is applied to a fatigue-sensitive detail of an existing steel bridge, and the monitoring planning is illustrated with a hull structure of a ship under fatigue.

6.2 Multi-Objective Probabilistic Optimum Inspection Planning

The optimum inspection planning can be based on multiple objectives considering the fatigue damage occurrence and propagation, damage detection, and effects of inspection and maintenance on service life and life-cycle cost (Ellingwood and Mori 1997; Madsen et al. 1991; Soliman and Frangopol 2014; Kim and Frangopol 2018a, 2018b). Multi-objective optimization for inspection planning results in multiple trade-off solutions (i.e., Pareto optimal set), which are associated with the inspection application times. In this section, six objectives (i.e., $O_{I,1}$ = maximizing the lifetime probability of fatigue crack damage detection P_{det} (see Eq. (3.12)); $O_{I,2}$ = minimizing the expected fatigue crack damage detection delay $E(t_{del})$ (see Eq. (3.19)); $O_{I,3}$ = minimizing the fatigue crack damage detection time-based probability of failure P_f (see Eq. (3.25)); $O_{I,4}$ = minimizing the expected maintenance delay $E(t_{mdl})$ (see Eq. (5.1)); $O_{I,5}$ = maximizing the expected extended service life $E(t_{life})$ (see Eq. (5.3)); $O_{I,6}$ = minimizing the expected life-cycle cost $E(C_{lcc})$ (see Eq. (5.5))) are used to establish the optimum inspection plans. As an illustrative example, a fatigue-sensitive detail of the I-64 Bridge (Connor and Fisher 2001) presented in Sub-section 5.2.4 is investigated. The optimization problems in this chapter are solved using the multi-objective genetic algorithm (MOGA) of MATLAB® version R2016b (MathWorks 2016). The Pareto optimal set is obtained after 1000 generations with 200 populations. In the illustrative example, the optimum inspection planning is based on nondestructive ultrasonic (UL) inspection.

6.2.1 *Bi-Objective Optimum Inspection Planning*

For optimum inspection planning, the bi-objective optimization is formulated based on three combinations of two objectives (i.e., $\{O_{I,1}, O_{I,2}\}$, $\{O_{I,3}, O_{I,4}\}$ and $\{O_{I,5}, O_{I,6}\}$) as

$$\text{Find } \mathbf{t_{insp}} = \{t_{insp,1}, t_{insp,2}, \ldots, t_{insp,Ninsp}\} \quad (6.1a)$$

$$\text{for } \{O_{I,1}, O_{I,2}\}, \{O_{I,3}, O_{I,4}\} \text{ or } \{O_{I,5}, O_{I,6}\} \quad (6.1b)$$

$$\text{such that } t_{insp,i} - t_{insp,i-1} \geq 1 \text{ year and } t_{insp,i} \leq 20 \text{ years} \quad (6.1c)$$

$$\text{given } N_{insp} \text{ and type of the inspection method} \quad (6.1d)$$

where $\mathbf{t_{insp}}$ = vector of design variables consisting of inspection times $t_{insp,i}$ (years). The time interval between inspections should be larger than 1 year, and the inspection time $t_{insp,i}$ should be less than 20 years (see Eq.

(6.1c)). The number of inspections N_{insp} and type of the inspection (i.e., UL inspection) are given as indicated in Eq. (6.1d).

Figure 6.1 shows the Pareto sets of the bi-objective optimization problem defined in Eq. (6.1) for the objective set $\{O_{I,1}, O_{I,2}\}$. Any point in Figure 6.1 can be a solution for the optimum inspection planning. It is worth noting that the Pareto sets in Figure 6.1 is associated with Figure 1.8(d). The inspection times and objective values of the representative optimum solutions in Figure 6.1 (i.e., $I_{b,1}$ to $I_{b,9}$) are provided in Table 6.1. If the solution $I_{b,1}$ in Figure 6.1(a) is selected for inspection planning, the UL inspection has to be applied at $t_{insp,1}$ = 13.22 years, and the lifetime probability of fatigue crack damage detection P_{det} and expected fatigue crack damage detection delay $E(t_{del})$ will be 0.814 and 10.35 years, respectively (see Table 6.1). The solution $I_{b,1}$ results in the smallest P_{det} and $E(t_{del})$ among the Pareto optimal set in Figure 6.1(a). Conversely, the solution $I_{b,3}$ leads to the largest P_{det} = 0.819 and $E(t_{del})$ = 10.47 years, and the associated UL inspection time is 14.21 years. Furthermore, the Pareto optimal set for two UL inspections are illustrated in Figure 6.1(b). The two UL inspections at 11.73 years and 16.27 years produce the largest P_{det} (i.e., 0.913) and $E(t_{del})$ (i.e., 7.51 years) as shown in Figure 6.1(b) and Table 6.1. If the three UL inspections are applied at 8.25 years, 11.28 years and 16.77 years (see the solution $I_{b,7}$ in Table 6.1), the smallest P_{det} (i.e., 0.934) and $E(t_{del})$ (i.e., 5.61 years) among the Pareto optimal set in Figure 6.1(c) can be obtained.

The Pareto optimal sets of the bi-objective optimization problem for the objective sets $\{O_{I,3}, O_{I,4}\}$ and $\{O_{I,5}, O_{I,6}\}$ are provided in Figure 6.2, where the number of inspections N_{insp} = 2 is given. The Pareto optimal sets for the objective sets $\{O_{I,3}, O_{I,4}\}$ and $\{O_{I,5}, O_{I,6}\}$ are associated with Figures 6.2(a) and 6.2(b), respectively. The inspection times and objective values of the representative optimum solutions $I_{b,10}$ to $I_{b,15}$ can be found in Table 6.1. The solution $I_{b,11}$ in Figure 6.2(a) results in the fatigue crack damage detection time-based probability of failure P_f = 0.080 and expected maintenance delay $E(t_{mdl})$ = 11.98 years. By selecting the solution $I_{b,12}$ instead of the solutions $I_{b,11}$, P_f increases by 30% (i.e., from 0.080 to 0.104), and $E(t_{mdl})$ decreases by 2.1% (i.e., from 11.98 years to 11.73 years). The two UL inspections at 11.28 years and 19.99 years associated with the solution $I_{b,14}$, lead to $E(t_{life})$ = 40.02 years and $E(C_{lcc})$ = \$238,813 as shown in Figure 6.2(b).

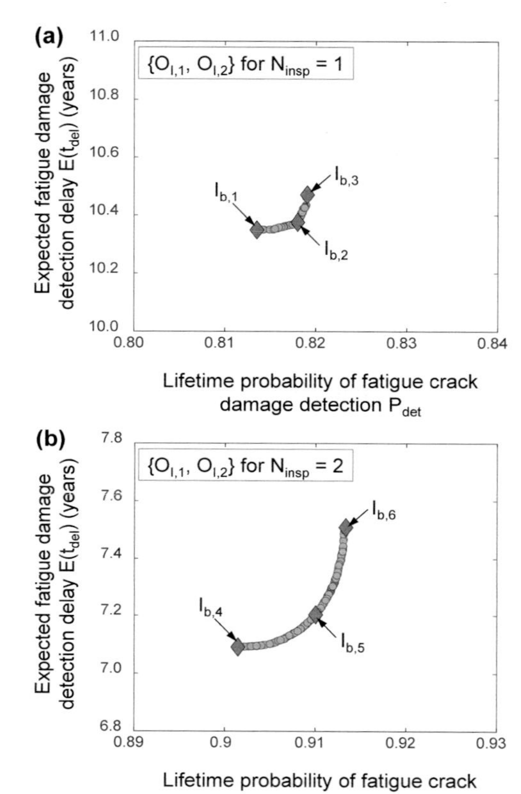

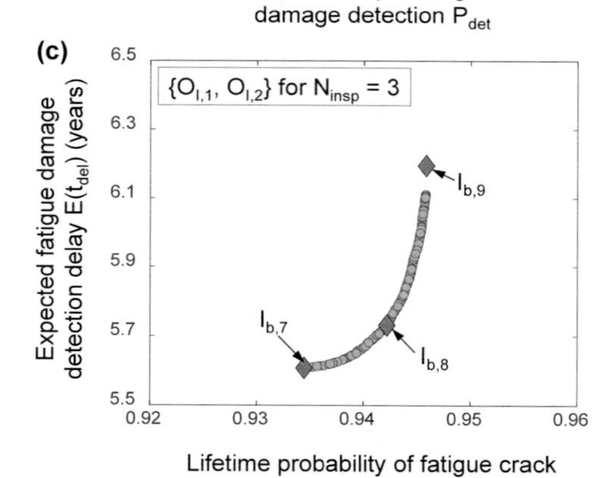

Figure 6.1 Pareto sets of bi-objective optimization associated with $\{O_{I,1}, O_{I,2}\}$: (a) $N_{insp} = 1$; (b) $N_{insp} = 2$; (c) $N_{insp} = 3$.

Table 6.1 Inspection times and associated objective values of the bi-objective optimum solutions in Figures 6.1 and 6.2.

Optimum solutions	Optimum inspection times (years)			Objective values						
	$t_{insp,1}$	$t_{insp,2}$	$t_{insp,3}$	P_{det}	$E(t_{del})$ (years)	P_f	$E(t_{mdl})$ (years)	$E(t_{life})$ (years)	$E(C_{lcc})$ (USD)
$I_{b,1}$	13.22	-	-	0.814	10.35	-	-	-	-
$I_{b,2}$	13.67	-	-	0.818	10.38	-	-	-	-
$I_{b,3}$	14.21	-	-	0.819	10.47	-	-	-	-
$I_{b,4}$	9.65	15.48	-	0.902	7.09	-	-	-	-
$I_{b,5}$	10.74	16.03	-	0.910	7.20	-	-	-	-
$I_{b,6}$	11.73	16.27	-	0.913	7.51	-	-	-	-
$I_{b,7}$	8.25	11.28	16.77	0.934	5.61	-	-	-	-
$I_{b,8}$	9.27	12.51	17.53	0.942	5.73	-	-	-	-
$I_{b,9}$	10.69	13.34	17.61	0.946	6.20	-	-	-	-
$I_{b,10}$	10.61	16.63	-	-	-	0.066	13.27	-	-
$I_{b,11}$	12.88	18.01	-	-	-	0.080	11.98	-	-
$I_{b,12}$	14.14	19.97	-	-	-	0.104	11.73	-	-
$I_{b,13}$	11.73	14.96	-	-	-	-	-	36.01	163732
$I_{b,14}$	11.28	19.99	-	-	-	-	-	40.02	238813
$I_{b,15}$	10.00	20.00	-	-	-	-	-	41.34	277117

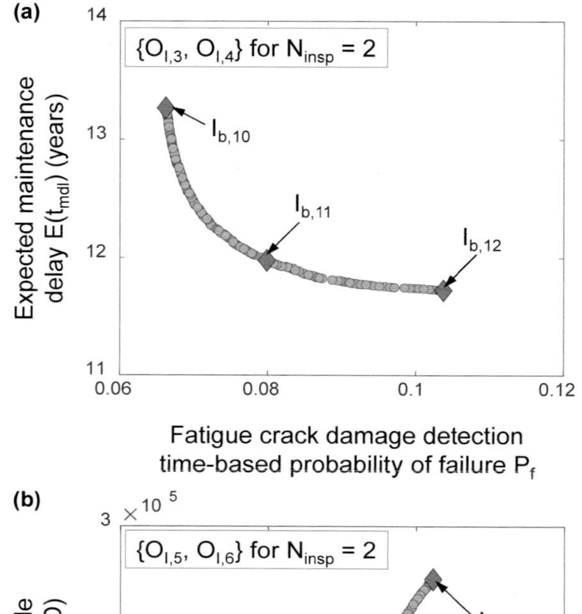

Figure 6.2 Pareto sets of bi-objective optimization for $N_{insp} = 2$: (a) $\{O_{I,3}, O_{I,4}\}$; (b) $\{O_{I,5}, O_{I,6}\}$.

6.2.2 Tri-Objective Optimum Inspection Planning

By considering the two objective sets $\{O_{I,1}, O_{I,2}, O_{I,3}\}$ and $\{O_{I,4}, O_{I,5}, O_{I,6}\}$, the tri-objective optimization for inspection planning is formulated as

Find $\mathbf{t_{insp}} = \{t_{insp,1}, t_{insp,2}, \ldots, t_{insp,Ninsp}\}$ (6.2a)

for $\{O_{I,1}, O_{I,2}, O_{I,3}\}$ or $\{O_{I,4}, O_{I,5}, O_{I,6}\}$ (6.2b)

The design variables are the inspection times $\mathbf{t_{insp}}$ (years). The constraints and given conditions of this tri-objective optimization problem are identical to those in the bi-objective optimization as indicated in Eqs. (6.1c) and (6.1d).

As shown in Figure 6.3, the Pareto optimal sets are obtained by solving the tri-objective optimization problem defined in Eq. (6.2), where the objective set $\{O_{I,1}, O_{I,2}, O_{I,3}\}$ is considered. Table 6.2 indicates the inspection times and objective values of the representative solutions in Figure 6.3 (i.e., $I_{t,1}$, $I_{t,2}$ and $I_{t,3}$). When the solution $I_{t,1}$ is selected from the Pareto set in Figure 6.3(a), the UL inspection has to be applied at 14.21 years, in order to obtain the largest lifetime probability of fatigue crack damage detection P_{det} of 0.819, as indicated in Table 6.2. If the smallest expected fatigue damage detection delay $E(t_{del})$ is needed when one solution is selected from the Pareto optimal set in Figure 6.3(b), the solution $I_{t,2}$ can be used as an inspection plan which requires two inspections at 9.65 years and 15.48 years, and leads to $P_{det} = 0.901$, $E(t_{del}) = 7.09$ years and $P_f = 0.069$. Furthermore, the solution $I_{t,3}$ in Figure 6.3(c) results in the smallest P_f among the Pareto optimal set for three inspections (i.e., $N_{insp} = 3$), and the associated inspection times and objective values are provided in Table 6.2.

Figure 6.4 shows the Pareto optimal sets of the tri-objective optimization problems with the objective set $\{O_{I,4}, O_{I,5}, O_{I,6}\}$. The solution $I_{t,4}$ produces the smallest expected maintenance delay $E(t_{mdl})$ (i.e., 12.11 years) among the Pareto optimal set in Figure 6.4(a), where the associated inspection time is 14.91 years as presented in Table 6.2. If the largest expected extended service life $E(t_{life})$ among the Pareto optimal set in Figure 6.4(b) is required, the solution $I_{t,5}$ can be selected. As indicated in Table 6.2, the solution $I_{t,6}$ associated with inspections at 11.15 years, 13.33 years and 15.82 years can lead to the smallest $E(C_{lcc})$ from the Pareto optimal set in Figure 6.4(c).

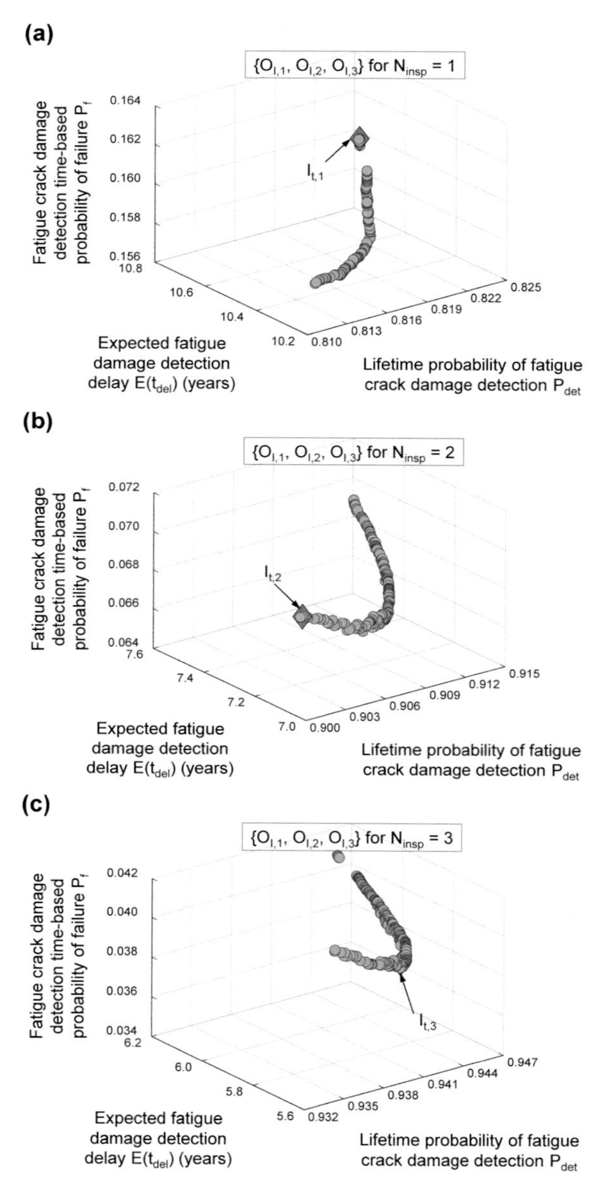

Figure 6.3 Pareto sets of tri-objective optimization with $\{O_{I,1}, O_{I,2}, O_{I,3}\}$: (a) $N_{insp} = 1$; (b) $N_{insp} = 2$; (c) $N_{insp} = 3$.

Table 6.2 Inspection times and associated objective values of the tri-objective optimum solutions in Figures 6.3 and 6.4.

Optimum solutions	**Optimum inspection times (years)**			**Objective values**					
	$t_{insp,1}$	$t_{insp,2}$	$t_{insp,3}$	P_{det}	$E(t_{del})$ (years)	P_f	$E(t_{mdl})$ (years)	$E(t_{life})$ (years)	$E(C_{lcc})$ (USD)
$I_{t,1}$	14.21	-	-	0.819	10.47	0.163	-	-	-
$I_{t,2}$	9.65	15.48	-	0.901	7.09	0.069	-	-	-
$I_{t,3}$	9.29	12.78	17.61	0.946	6.12	0.040	-	-	-
$I_{t,4}$	14.91	-	-	-	-	-	12.11	37.10	189614
$I_{t,5}$	10.00	20.00	-	-	-	-	13.65	41.34	277117
$I_{t,6}$	11.15	13.33	15.82	-	-	-	13.04	35.88	164651

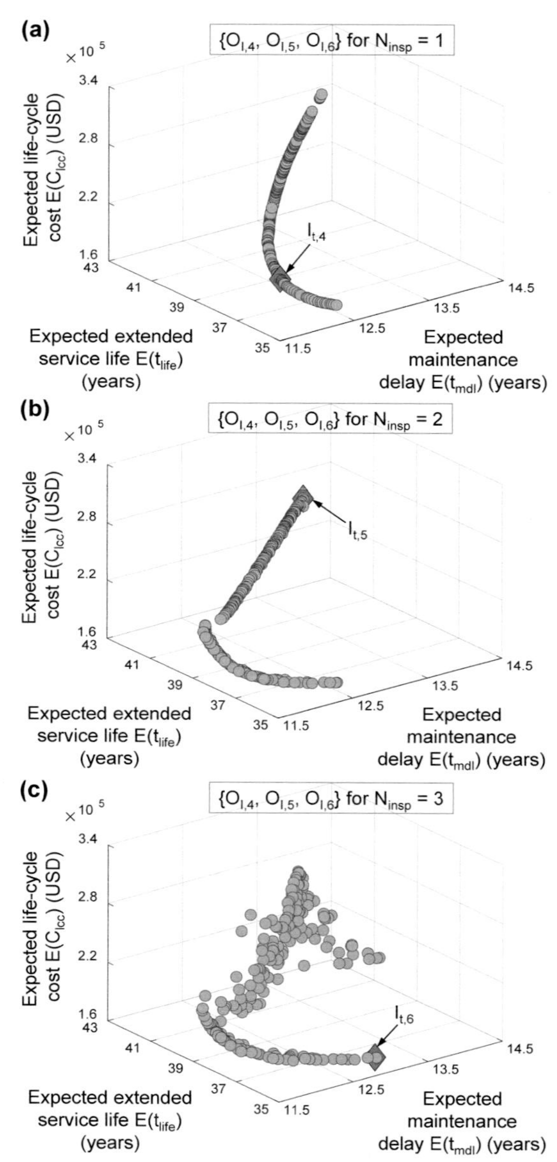

Figure 6.4 Pareto sets of tri-objective optimization with $\{O_{I,4}, O_{I,5}, O_{I,6}\}$: (a) $N_{insp} = 1$; (b) $N_{insp} = 2$; (c) $N_{insp} = 3$.

6.2.3 Quad-Objective Optimum Inspection Planning

The two objective sets $\{O_{I,1}, O_{I,2}, O_{I,3}, O_{I,4}\}$ and $\{O_{I,3}, O_{I,4}, O_{I,5}, O_{I,6}\}$ are considered for the quad-objective optimization as

$$\text{Find } \mathbf{t_{insp}} = \{t_{insp,1}, t_{insp,2}, \ldots, t_{insp,Ninsp}\} \tag{6.3a}$$

$$\text{for } \{O_{I,1}, O_{I,2}, O_{I,3}, O_{I,4}\} \text{ or } \{O_{I,3}, O_{I,4}, O_{I,5}, O_{I,6}\} \tag{6.3b}$$

The same constraints and given conditions used in the bi- and tri-objective optimizations are applied to this quad-objective optimization. The number of UL inspections is assumed to be two (i.e., $N_{insp} = 2$). The Pareto solutions by solving the quad-objective optimizations with $\{O_{I,1}, O_{I,2}, O_{I,3}, O_{I,4}\}$ and $\{O_{I,3}, O_{I,4}, O_{I,5}, O_{I,6}\}$ are presented in Figures 6.5 and 6.6, respectively. The values of design variables and objective functions of four representative solutions (i.e., $I_{q,1}$, $I_{q,2}$, $I_{q,3}$ and $I_{q,4}$) are provided in Table 6.3.

Since the dimension of the Pareto solutions is equal to the number of objectives to be considered (i.e., four), the Pareto solutions for $\{O_{I,1}, O_{I,2}, O_{I,3}, O_{I,4}\}$ are illustrated in the 3D Cartesian coordinate system and the parallel coordinate system with four vertical axes, as shown in Figures 6.5(a) and 6.5(b). Each point in the 3D Cartesian coordinate system is represented by the polyline connecting the corresponding values on the vertical axes of the parallel coordinate system. For example, the point for the solution $I_{q,1}$ in Figure 6.5(a) is associated with the polyline for the solution $I_{q,1}$ in Figure 6.5(b). The solution $I_{q,1}$ is associated with the two inspections at 9.65 years and 15.43 years (see Table 6.3), and leads to the smallest expected fatigue damage detection delay $E(t_{del})$ from all the Pareto solutions (see Figure 6.5). The Pareto solution with the smallest expected maintenance delay $E(t_{mdl})$ corresponds to the solution $I_{q,2}$, which can be obtained by applying two inspections at 14.14 years and 19.94 years (see Figure 6.5 and Table 6.3). From the Pareto solutions for $\{O_{I,3}, O_{I,4}, O_{I,5}, O_{I,6}\}$ in Figure 6.6, the solutions $I_{q,3}$ and $I_{q,4}$ produce the smallest expected maintenance delay $E(t_{mdl})$ and the smallest expected life-cycle cost $E(C_{lcc})$, respectively.

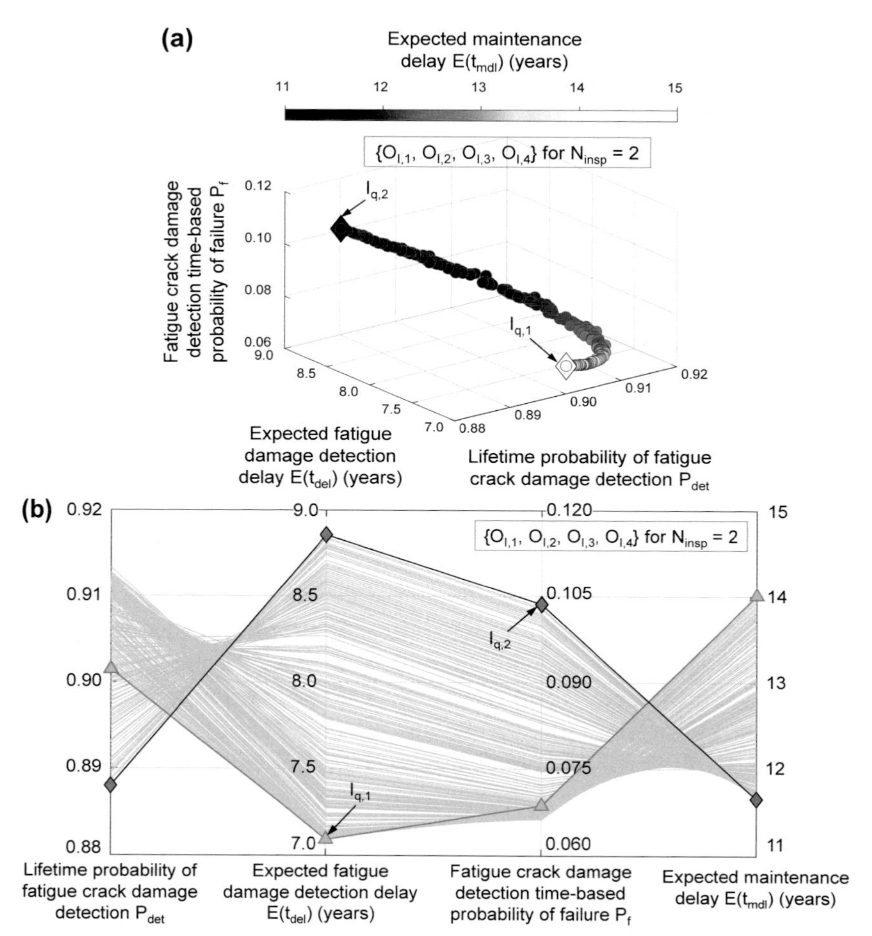

Figure 6.5 Pareto solutions of quad-objective optimization with $\{O_{I,1}, O_{I,2}, O_{I,3}, O_{I,4}\}$ for $N_{insp} = 2$: (a) 3D Cartesian coordinate system; (b) parallel coordinate system.

Color version at the end of the book

6.2.4 *Six-Objective Optimum Inspection Planning*

When all the six objectives $O_{I,1}$, $O_{I,2}$, $O_{I,3}$, $O_{I,4}$, $O_{I,5}$ and $O_{I,6}$ are considered simultaneously for optimum inspection planning, the Pareto solutions are illustrated in the parallel coordinate system as shown in Figure 6.7. The

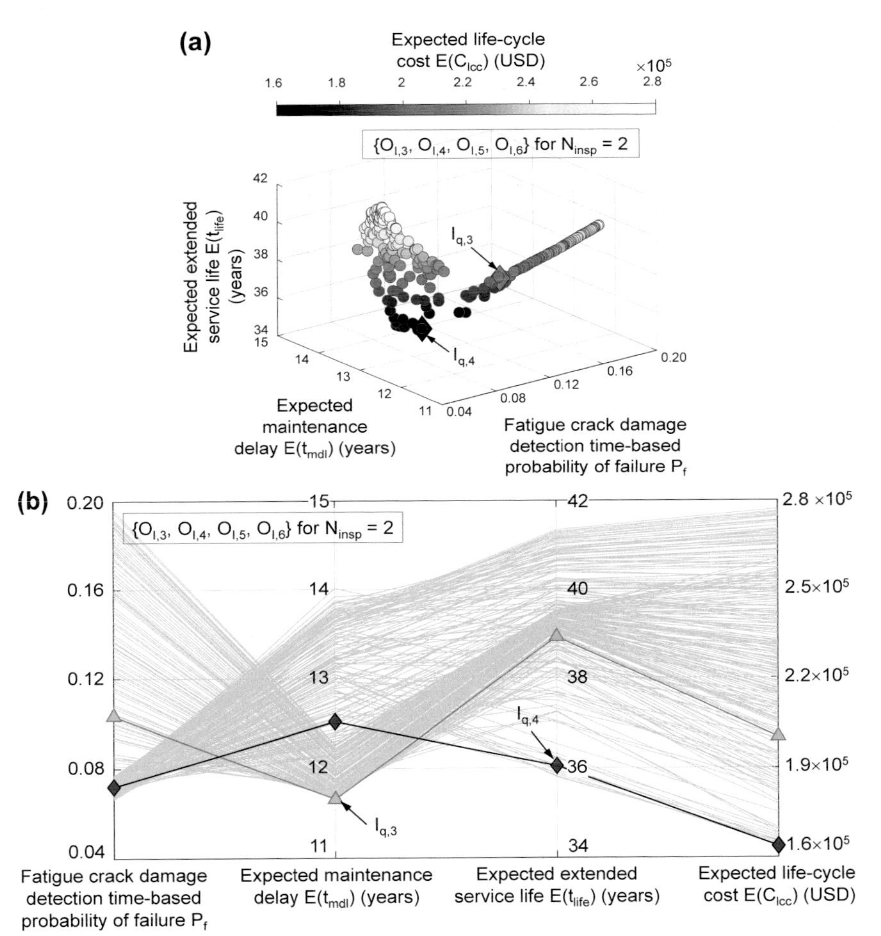

Figure 6.6 Pareto solutions of quad-objective optimization with $\{O_{I,3}, O_{I,4}, O_{I,5}, O_{I,6}\}$ for $N_{insp} = 2$: (a) 3D Cartesian coordinate system; (b) parallel coordinate system.

Color version at the end of the book

solutions $I_{s,1}$, $I_{s,2}$ and $I_{s,3}$ among the Pareto solutions are associated with the smallest $E(t_{del})$, $E(t_{mdl})$ and $E(C_{lcc})$, respectively. The inspection times $t_{insp,1}$ and $t_{insp,2}$ for these three solutions are indicated in Table 6.3. For example, if the smallest $E(C_{lcc})$ is required for the optimum inspection planning considering the six objectives $O_{I,1}$, $O_{I,2}$, $O_{I,3}$, $O_{I,4}$, $O_{I,5}$ and $O_{I,6}$, the inspections should be applied at 11.83 years and 14.96 years (see Table 6.3).

Table 6.3 Inspection times and associated objective values of the multi-objective optimum solutions in Figures 6.5, 6.6 and 6.7.

Optimum solutions	Optimum inspection times (years)		Objective values					
	$t_{insp,1}$	$t_{insp,2}$	P_{det}	$E(t_{del})$ (years)	P_f	$E(t_{mdl})$ (years)	$E(t_{life})$ (years)	$E(C_{lcc})$ (USD)
$I_{q,1}$	9.65	15.43	0.901	7.09	0.069	14.01	-	-
$I_{q,2}$	14.14	19.94	0.888	8.86	0.104	11.65	-	-
$I_{q,3}$	14.14	19.81	-	-	0.103	11.65	38.93	200353
$I_{q,4}$	11.83	14.96	-	-	0.0718	12.52	36.05	163743
$I_{s,1}$	9.66	15.48	0.902	7.09	0.069	14.00	36.28	183912
$I_{s,2}$	14.15	19.94	0.888	8.87	0.104	11.65	38.98	201938
$I_{s,3}$	11.83	14.96	0.911	7.64	0.072	12.52	36.05	163744

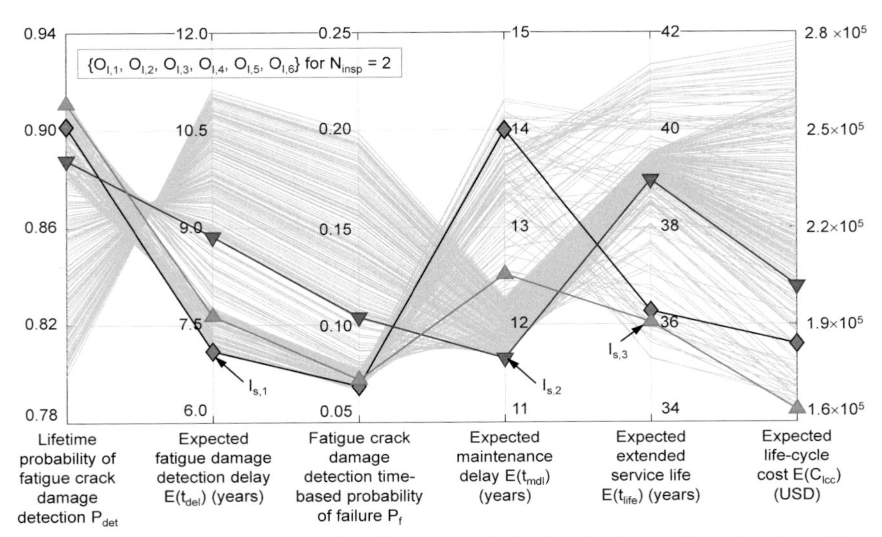

Figure 6.7 Pareto solutions of six-objective optimization for inspection planning with $N_{insp} = 2$ in the parallel coordinate system.

6.3 Multi-Objective Probabilistic Optimum Monitoring Planning

The optimum monitoring planning can be based on five objectives: $O_{M,1}$ = minimizing the expected fatigue crack damage detection delay $E(t_{del})$ (see Eq. (3.20)); $O_{M,2}$ = minimizing the fatigue crack damage detection time-based probability of failure P_f (see Eq. (3.25)); $O_{M,3}$ = minimizing the expected maintenance delay $E(t_{mdl})$ (see Eq. (5.8)); $O_{M,4}$ = maximizing the expected extended service life $E(t_{life})$ (see Eq. (5.10)); $O_{M,5}$ = minimizing the expected life-cycle cost $E(C_{lcc})$ (see Eq. (5.11)). The monitoring planning presented herein is illustrated with a ship hull structure under fatigue. The detailed information related to probabilistic fatigue crack propagation can be found in Sub-section 3.4.4 and Table 3.1. The estimation of the life-cycle cost for monitoring planning is based on the cost indicated in Table 5.3. The MOGA of MATLAB version R2016b (MathWorks 2016) is used to solve the multi-objective optimization for monitoring planning, where 1000 generations with 200 populations are applied.

6.3.1 Bi-Objective Optimum Monitoring Planning

The formulation of the bi-objective optimization for monitoring planning is expressed as

$$\text{Find } \mathbf{t_{ms}} = \{t_{ms,1}, t_{ms,2}, \ldots, t_{ms,Nmon}\} \quad (6.4a)$$

$$\text{for } \{O_{M,1}, O_{M,2}\}, \{O_{M,3}, O_{M,4}\} \text{ or } \{O_{M,4}, O_{M,5}\} \quad (6.4b)$$

$$\text{such that 1 year} \leq t_{ms,i} - (t_{ms,i-1} + t_{md}) \leq 20 \text{ years} \quad (6.4c)$$

$$\text{given } N_{mon} = 2 \text{ and } t_{md} = 0.3 \text{ years} \quad (6.4d)$$

The design variables are the monitoring starting times $\mathbf{t_{ms}}$ (years) (see Eq. (6.4a)). The three objective sets $\{O_{M,1}, O_{M,2}\}$, $\{O_{M,3}, O_{M,4}\}$ and $\{O_{M,4}, O_{M,5}\}$ are considered for the bi-objective optimization (see Eq. (6.4b)). The non-monitoring time interval (i.e., $t_{ms,i} - (t_{ms,i-1} + t_{md})$) has to be larger than 1 year and less than 20 years (see Eq. (6.4c)). The given number of monitorings N_{mon} and monitoring duration t_{md} are 2 and 0.3 year, respectively (see Eq. (6.4d)).

The Pareto sets for the objective sets $\{O_{M,1}, O_{M,2}\}$, $\{O_{M,3}, O_{M,4}\}$ and $\{O_{M,4}, O_{M,5}\}$ are illustrated in Figure 6.8, by solving the bi-objective optimization problem defined in Eq. (6.4). The design variables (i.e., monitoring starting times $t_{ms,1}$ and $t_{ms,2}$) and objective values of the solutions $M_{b,1}$ to $M_{b,6}$ in Figure 6.8 are provided in Table 6.4. When the smallest expected fatigue damage detection delay $E(t_{del})$ is required among the Pareto optimal set in Figure 6.8(a), the solution $M_{b,1}$ is selected, and the monitorings have to be applied at 4.64 years and 10.10 years. The solution $M_{b,1}$ also results in the largest fatigue crack damage detection time-based probability of failure P_f. On the other hand, the solution $M_{b,2}$ leads to the smallest P_f and the largest $E(t_{del})$.

As shown in Figure 6.8(b), the solutions $M_{b,3}$ is associated with the smallest expected maintenance delay $E(t_{mdl})$ and the smallest expected extended service life $E(t_{life})$, when the two objectives $O_{M,3}$ and $O_{M,4}$ are considered for the bi-objective optimization. By applying the two monitorings at 10.41 years and 21.11 years, the largest $E(t_{mdl})$ and

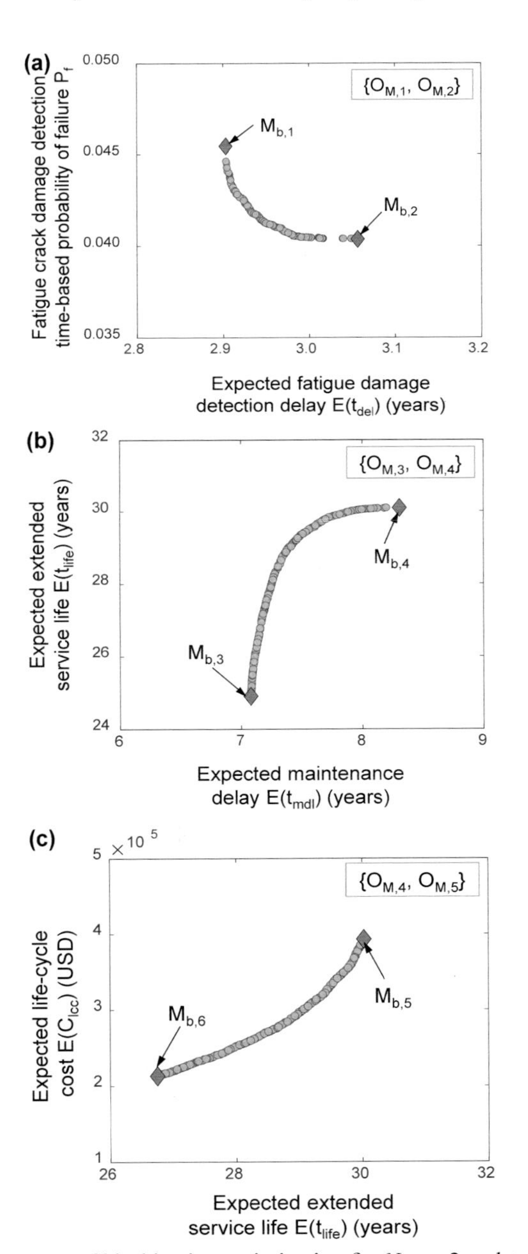

Figure 6.8 Pareto sets of bi-objective optimization for $N_{mon} = 2$ and $t_{md} = 0.3$ year: (a) $\{O_{M,1}, O_{M,2}\}$; (b) $\{O_{M,3}, O_{M,4}\}$; (c) $\{O_{M,4}, O_{M,5}\}$.

Table 6.4 Monitoring starting times and associated objective values of the bi-objective optimum solutions in Figures 6.8.

Optimum solutions	Optimum monitoring starting times (years)		Objective values				
	$t_{ms,1}$	$t_{ms,2}$	$E(t_{del})$ (years)	P_f	$E(t_{mdl})$ (years)	$E(t_{life})$ (years)	$E(C_{lcc})$ (USD)
$M_{b,1}$	4.64	10.10	2.90	0.046	-	-	-
$M_{b,2}$	3.75	8.70	3.06	0.0404	-	-	-
$M_{b,3}$	7.38	12.94	-	-	7.08	24.90	-
$M_{b,4}$	10.41	21.11	-	-	8.31	30.09	-
$M_{b,5}$	9.69	19.69	-	-	-	30.03	392835
$M_{b,6}$	6.37	13.03	-	-	-	26.75	213132

the largest $E(t_{life})$ from the Pareto optimal set in Figure 6.8(b) can be obtained, which is represented by the solution $M_{b,4}$. Furthermore, if the solution $M_{b,5}$ in Figure 6.8(c) is selected for the monitoring planning, the monitoring starting times should be 9.69 years and 19.69 years, and $E(t_{life})$ and $E(C_{lcc})$ will be 30.03 years and \$392,835 (see Table 6.4), which are the largest values of $E(t_{life})$ and $E(C_{lcc})$ from the Pareto set in Figure 6.8(c).

6.3.2 *Tri-Objective Optimum Monitoring Planning*

The tri-objective optimum monitoring planning is based on the objective sets $\{O_{M,1}, O_{M,2}, O_{M,3}\}$ and $\{O_{M,3}, O_{M,4}, O_{M,5}\}$, which is formulated as

$$\text{Find } \mathbf{t_{ms}} = \{t_{ms,1}, t_{ms,2}, \ldots, t_{ms,Nmon}\} \tag{6.5a}$$

$$\text{for } \{O_{M,1}, O_{M,2}, O_{M,3}\} \text{ or } \{O_{M,3}, O_{M,4}, O_{M,5}\} \tag{6.5b}$$

The constraints and given conditions in Eq. (6.4) are applied for this tri-objective optimization problem. Figure 6.9 shows the Pareto sets for $\{O_{M,1}, O_{M,2}, O_{M,3}\}$ and $\{O_{M,3}, O_{M,4}, O_{M,5}\}$. The solutions $M_{t,1}$, $M_{t,2}$ and $M_{t,3}$ are associated with the smallest $E(t_{del})$, P_f and $E(t_{mdl})$, respectively, from the Pareto set in Figure 6.9(a). The design variables and objective values for these three solutions are provided in Table 6.5. Furthermore, the monitoring plans represented by the solutions $M_{t,4}$, $M_{t,5}$ and $M_{t,6}$ lead to the smallest $E(t_{mdl})$, the largest $E(t_{life})$ and the smallest $E(C_{lcc})$ from the Pareto set in Figure 6.9(b). It should be noted that the monitoring starting times at 7.38 years and 12.94 years result in the smallest $E(t_{mdl})$ of 7.08 years from the Pareto sets for both objective sets $\{O_{M,1}, O_{M,2}, O_{M,3}\}$ and $\{O_{M,3}, O_{M,4}, O_{M,5}\}$ (see the solutions $M_{t,3}$ and $M_{t,4}$ in Figure 6.9 and Table 6.5).

6.3.3 *Quad-Objective Optimum Monitoring Planning*

Considering the two objective sets $\{O_{M,1}, O_{M,2}, O_{M,3}, O_{M,4}\}$ and $\{O_{M,2}, O_{M,3}, O_{M,4}, O_{M,5}\}$, the quad-objective optimization for monitoring planning is formulated as

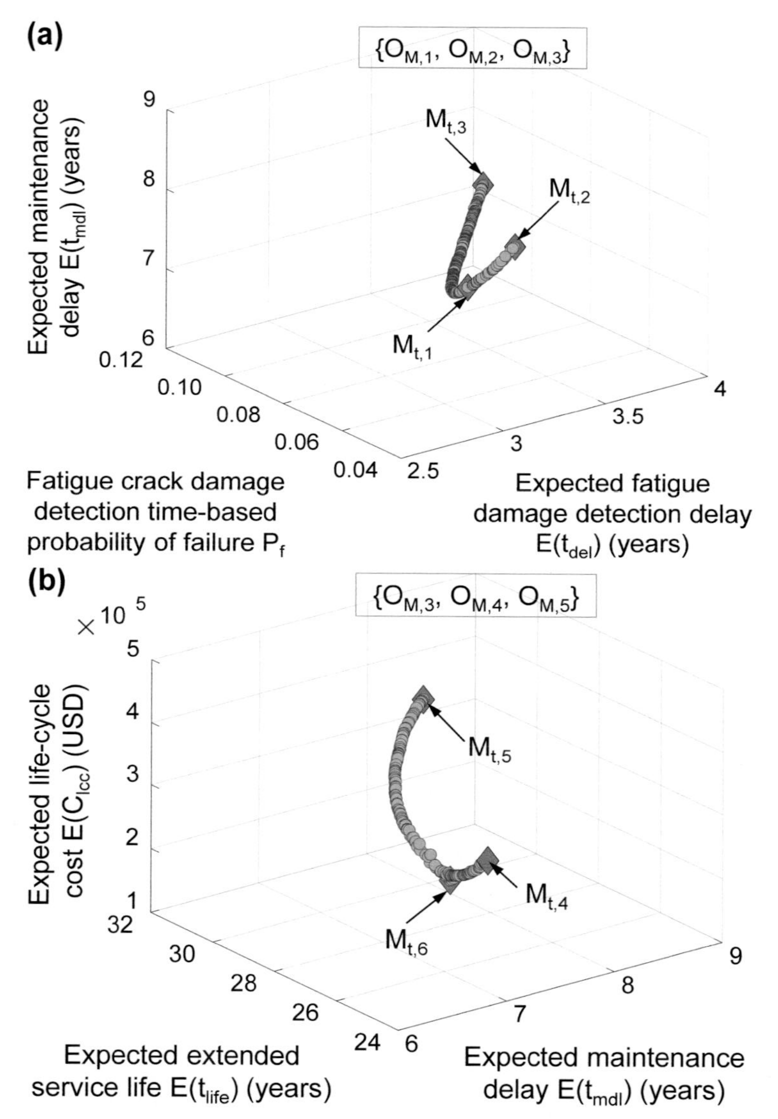

Figure 6.9 Pareto sets of tri-objective optimization for $N_{mon} = 2$ and $t_{md} = 0.3$ year: (a) $\{O_{M,1}, O_{M,2}, O_{M,3}\}$; (b) $\{O_{M,3}, O_{M,4}, O_{M,5}\}$.

Table 6.5 Monitoring starting times and associated objective values of the tri-objective optimum solutions in Figures 6.9.

Optimum solutions	Optimum monitoring starting times (years)		Objective values				
	$t_{ms,1}$	$t_{ms,2}$	$E(t_{del})$ (years)	P_f	$E(t_{mdl})$ (years)	$E(t_{life})$ (years)	$E(C_{lcc})$ (USD)
$M_{t,1}$	4.64	10.10	2.90	0.045	7.78	-	-
$M_{t,2}$	3.75	8.67	3.06	0.040	8.27	-	-
$M_{t,3}$	7.38	12.94	4.00	0.116	7.08	-	-
$M_{t,4}$	7.38	12.94	-	-	7.08	24.89	295646
$M_{t,5}$	9.69	19.68	-	-	7.95	30.02	392456
$M_{t,6}$	6.37	13.03	-	-	7.27	26.75	213132

Find $\mathbf{t_{ms}} = \{t_{ms,1}, t_{ms,2}, \ldots, t_{ms,Nmon}\}$ (6.6a)

for $\{O_{M,1}, O_{M,2}, O_{M,3}, O_{M,4}\}$ or $\{O_{M,2}, O_{M,3}, O_{M,4}, O_{M,5}\}$ (6.6b)

The constraints and given conditions indicated in Eq. (6.4) are applied to find the monitoring starting times $\mathbf{t_{ms}}$ of this quad-objective optimization. The Pareto optimal solutions for $\{O_{M,1}, O_{M,2}, O_{M,3}, O_{M,4}\}$ and $\{O_{M,2}, O_{M,3}, O_{M,4}, O_{M,5}\}$ are illustrated in the parallel coordinate system in Figure 6.10. When the four objectives $O_{M,1}$, $O_{M,2}$, $O_{M,3}$ and $O_{M,4}$ are considered simultaneously, the solutions $M_{q,1}$, $M_{q,2}$, $M_{q,3}$ and $M_{q,4}$ are selected to obtain the smallest $E(t_{del})$, P_f, $E(t_{mdl})$ and the largest $E(t_{life})$, respectively, from the Pareto solutions (see Figure 6.10(a)). The solutions $M_{q,5}$, $M_{q,6}$, $M_{q,7}$ and $M_{q,8}$ in Figure 6.10(b) lead to the smallest P_f and $E(t_{mdl})$, the largest $E(t_{life})$, and the smallest $E(C_{lcc})$, respectively. The monitoring starting times and objective values for the solutions $M_{q,1}$ to $M_{q,8}$ can be found in Table 6.6.

It should be noted that the solution $M_{q,2}$ results in the smallest P_f, which is the best for the objective $O_{M,2}$, but produces the smallest $E(t_{life})$, which is the worst for the objective $O_{M,4}$ among the Pareto solutions in Figure 6.10(a). The solution $M_{q,4}$ resulting in the largest $E(t_{life})$ is the best solution for the objective $O_{M,4}$. However, the solution $M_{q,4}$ is the worst solution for the objectives $O_{M,1}$, $O_{M,2}$, $O_{M,3}$, since the largest $E(t_{del})$, P_f, $E(t_{mdl})$ are obtained, as shown in Figure 6.10(a). Moreover, if the four objectives $\{O_{M,2}, O_{M,3}, O_{M,4}, O_{M,5}\}$ are used for the quad-objective optimum monitoring planning, two monitorings at 3.75 years and 8.67 years (i.e., solution $M_{q,5}$ in Table 6.6) lead to the smallest P_f, and are the best solution only for the objective $O_{M,2}$. However, this solution is the worst solution for the objectives $O_{M,4}$ and $O_{M,5}$, as shown in Figure 6.10. Figure 6.11 provides the Pareto solutions in the parallel coordinate system, when the five objectives $O_{M,1}$, $O_{M,2}$, $O_{M,3}$, $O_{M,4}$ and $O_{M,5}$ are considered simultaneously for $N_{mon} = 2$ and $t_{md} = 0.3$ year. In Chapter 7, the decision making processes are addressed in order to select well-balanced solutions for optimum inspection and monitoring planning when multiple objectives are taken into account.

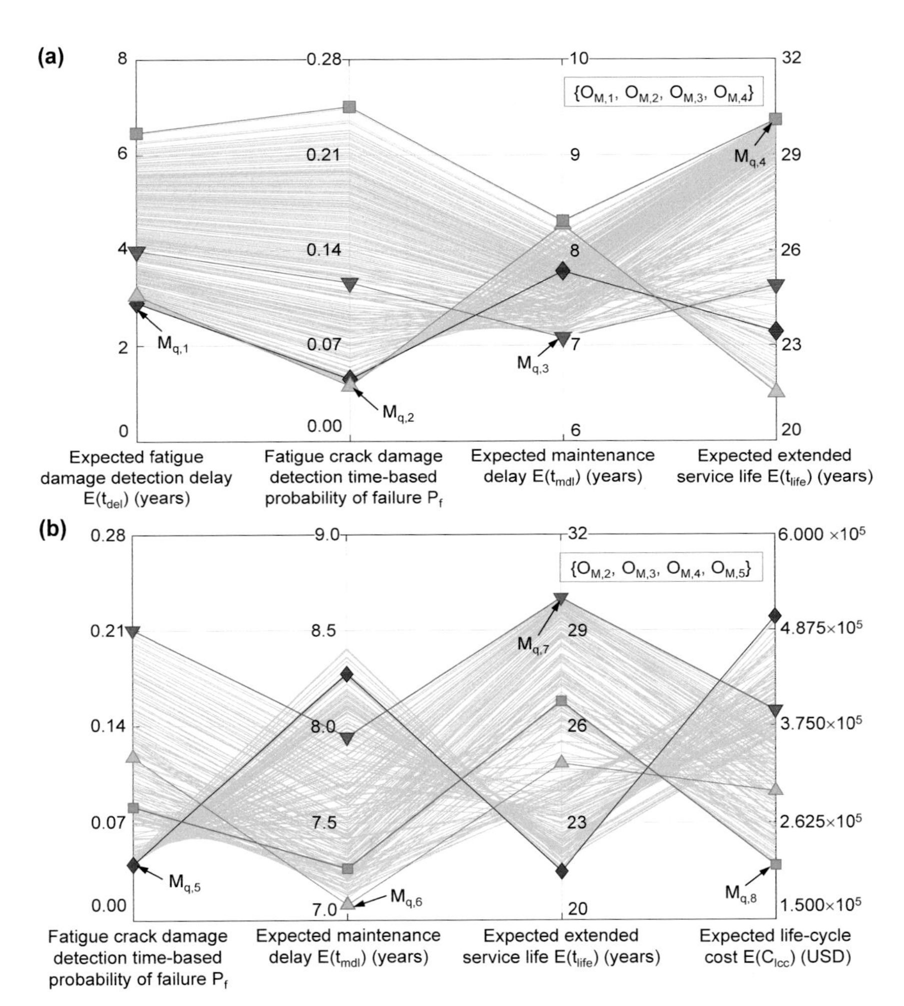

Figure 6.10 Pareto solutions of quad-objective optimization for $N_{mon} = 2$ and $t_{md} = 0.3$ year in the parallel coordinate system: (a) $\{O_{M,1}, O_{M,2}, O_{M,3}, O_{M,4}\}$; (b) $\{O_{M,2}, O_{M,3}, O_{M,4}, O_{M,5}\}$.

Table 6.6 Monitoring starting times and associated objective values of the tri-objective optimum solutions in Figures 6.10.

Optimum solutions	Optimum monitoring starting times (years)		Objective values				
	$t_{ms,1}$	$t_{ms,2}$	$E(t_{del})$ (years)	P_f	$E(t_{mdl})$ (years)	$E(t_{life})$ (years)	$E(C_{lcc})$ (USD)
$M_{q,1}$	4.64	10.10	2.90	0.045	7.78	23.42	-
$M_{q,2}$	3.75	8.67	3.06	0.040	8.27	21.50	-
$M_{q,3}$	7.38	12.94	4.00	0.116	7.08	24.89	-
$M_{q,4}$	10.40	21.10	6.47	0.245	8.31	30.09	-
$M_{q,5}$	3.75	8.67	-	0.040	8.27	21.50	503018
$M_{q,6}$	7.41	12.94	-	0.118	7.08	24.84	298443
$M_{q,7}$	9.66	19.64	-	0.211	7.94	30.01	392424
$M_{q,8}$	6.37	13.03	-	0.082	7.27	26.75	213132

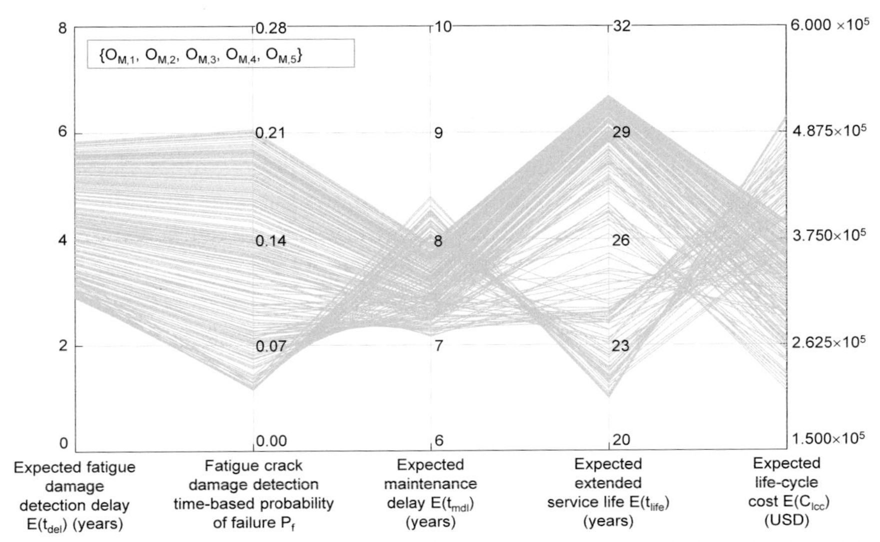

Figure 6.11 Pareto solutions of five-objective optimization for monitoring planning with $N_{mon} = 2$ and $t_{md} = 0.3$ year in the parallel coordinate system.

6.4 Conclusions

This chapter presents the multi-objective optimization approach for inspection and monitoring planning. The six objectives $O_{I,1}$ = maximizing the lifetime probability of fatigue crack damage detection P_{det}, $O_{I,2}$ = minimizing the expected fatigue crack damage detection delay $E(t_{del})$, $O_{I,3}$ = minimizing the fatigue crack damage detection time-based probability of failure P_f, $O_{I,4}$ = minimizing the expected maintenance delay $E(t_{mdl})$, $O_{I,5}$ = maximizing the expected extended service life $E(t_{life})$ and $O_{I,6}$ = minimizing the expected life-cycle cost $E(C_{lcc})$ are used to establish the inspection plans. The monitoring planning is based on the five objectives $O_{M,1}$ = minimizing $E(t_{del})$, $O_{M,2}$ = minimizing P_f, $O_{M,3}$ = minimizing $E(t_{mdl})$, $O_{M,4}$ = maximizing $E(t_{life})$ and $O_{M,6}$ = minimizing $E(C_{lcc})$. With these objectives, the bi-, tri-, and quad-objective optimization problems are investigated. Furthermore, the six objectives for optimum inspection planning and five objectives for optimum monitoring planning are also used simultaneously to formulate the multi-objective optimization problem.

By solving the multi-objective optimization, the Pareto optimal set is obtained. According to an increase in the number of objectives, the dimensionality of the Pareto optimal set increases. Therefore, it is difficult to illustrate the Pareto solutions associated with more than four objectives in the 3D Cartesian coordinate system. The parallel coordinate system is appropriate to plot the Pareto solutions based on more than four objectives. Practically, only one solution from the Pareto optimal set is required for inspection and/or monitoring planning of a single fatigue-sensitive detail. For this reason, the decision making process is required to select the best-balanced solution. Such a decision making framework is presented in Chapter 7.

Chapter 7

Decision Making for Multi-Objective Life-Cycle Optimization

CONTENTS

ABSTRACT

Chapter 7 presents a multi-objective decision making framework to deal with a large number of objectives efficiently and to select the best single optimum inspection and monitoring plan for practical applications. In this framework, there are two decision alternatives: decision making before and after solving multi-objective life-cycle optimization (MOLCO). The decision making before solving MOLCO results in a single optimum inspection and monitoring plan by integrating multi-objectives into a single objective. This integration requires estimating the weights of the objectives using a subjective weight determination method (e.g., ranking, rating and paired comparison methods). The decision making after MOLCO is based on the Pareto optimal solutions obtained from MOLCO. With the Pareto solution set, essential and redundant objectives are identified to improve the efficiency of the decision making. The weight factors of the essential objectives are computed using an objective weight determination method. Multiple attribute decision making (MADM) results in the best optimum inspection and monitoring plan.

7.1 Introduction

The development of novel probabilistic concepts and methods for optimum inspection, monitoring and maintenance planning has led to an increasing demand of objectives to be considered in the optimization process (Frangopol 2011; Frangopol and Soliman 2016; Kim and Frangopol 2017). Multi-objective optimization has been treated as an effective tool for integrating multiple objectives and for providing multiple well-balanced solutions. However, an increase in the number of objectives generally requires additional efforts for computation in order to find the entire Pareto optimal solutions, visualization of these solutions, and decision making to select the best Pareto optimal solution (Deb and Saxena 2006; Saxena et al. 2013; Verel et al. 2011). Therefore, it is necessary to develop a decision making framework for optimum inspection and monitoring planning, to deal with a large number of objectives efficiently and to select the best single optimum inspection and monitoring plan for practical applications (Brockhoff and Zitzler 2009; Kim and Frangopol 2018a, 2018b).

In this chapter, such a multi-objective decision making framework for fatigue-sensitive structures is presented. In this framework, the objective functions including the lifetime probability of fatigue crack damage detection P_{det}, expected damage detection delay $E(t_{del})$, damage detection time-based probability of failure P_f, expected maintenance delay $E(t_{mdl})$, expected extended service life $E(t_{life})$, and expected life-cycle cost $E(C_{lcc})$ are considered. The flowchart of the presented decision making framework is illustrated in Figure 7.1, where there are two decision alternatives: decision making before and after solving multi-objective life-cycle optimization (MOLCO). The decision making before solving MOLCO results in a single optimum inspection and monitoring plan by integrating the objectives used in multi-objective optimization into a single objective. This integration requires estimating the weights of the objectives using a subjective weight determination method (e.g., ranking, rating and paired comparison methods), as shown in Figure 7.1. The decision making after MOLCO is based on the Pareto optimal solutions obtained from MOLCO. With the Pareto solution set, essential and redundant objectives are identified to improve the efficiency of the decision making. The weight factors of the essential objectives are computed using an objective weight determination method such as the standard deviation (SD) method (Deng et al. 2000), criteria importance through the inter-criteria correlation (CRITIC) method (Diakoulaki et al. 1995), and correlation coefficient and standard deviation (CCSD) method (Wang and Luo 2010). Multiple attribute decision making (MADM) results in the best single optimum inspection and monitoring plan as shown in Figure 7.1.

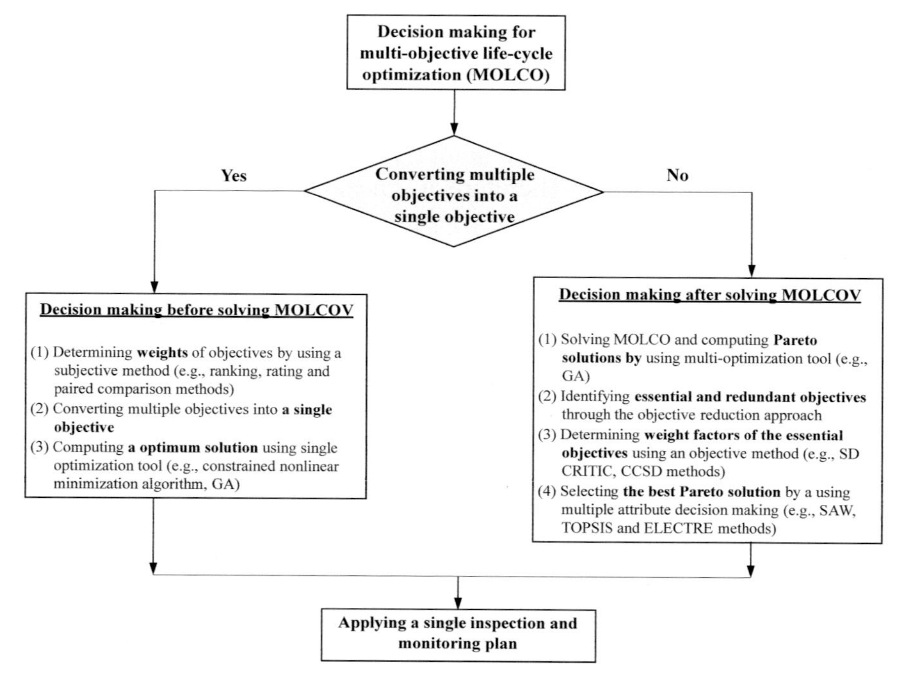

Figure 7.1 Decision making framework for multi-objective life-cycle optimization.

7.2 Decision Making Before Multi-Objective Life-Cycle Optimization

The main purpose of the decision making before solving MOLCO is to determine the weights of multiple objectives, and to convert a multi-objective optimization problem into a single-objective optimization problem for the life-cycle management of fatigue-sensitive structures. Using the weight sum method, the multiple objective functions to be minimized (or maximized) are converted into a single objective function $f_s(\mathbf{x})$ as (Arora 2016)

$$f_s(\mathbf{x}) = \sum_{i=1}^{N_{obj}} w_i f_i^{norm}(\mathbf{x}) \tag{7.1}$$

where **x** is a vector consisting of design variables, w_i is the weight factor of the ith normalized objective function $f_i^{norm}(\mathbf{x})$ satisfying $w_i \geq 0$ and $\sum_{i=1}^{N_{obj}} w_i = 1$, and N_{obj} is the number of objectives to be considered in MOLCO. The normalized objective function $f_i^{norm}(\mathbf{x})$ is expressed as

$$f_i^{norm}(\mathbf{x}) = \frac{f_i(\mathbf{x}) - f_i^{\min}}{f_i^{\max} - f_i^{\min}} \tag{7.2}$$

where $f_i(\mathbf{x})$ is the ith objective function, and f_i^{min} and f_i^{max} are the minimum and maximum f_i values for all **x** in the feasible design space, respectively. It should be noted that the weight sum method cannot result in the complete Pareto optimal set, but can be used extensively notwithstanding the large number of objectives to be considered simultaneously.

7.2.1 *Objective Weighting*

The weight of objective w_i in Eq. (7.1) can be determined using the subjective weight determination method. This method is based on the preference of the decision makers, when the information on objective values (e.g., correlation among the values of the objectives, and distribution of the objective values of the Pareto optimal solution set) is not available (Marler and Arora 2010). There are several representative subjective weight determination methods such as the ranking method, rating method and paired comparison method.

The ranking method is the simplest way to assess the weights of the objectives. The rank of 1 is assigned to the most important objective, and the least important objective has the rank N_{obj}, when the number of objectives for MOLCO is N_{obj}. The weight factor w_i of the ith objective function f_i can be expressed in two different ways (Stillwell et al. 1981)

$$w_i = \frac{\frac{1}{r_i}}{\sum_{k=1}^{N_{obj}} \left(\frac{1}{r_k} \right)} \tag{7.3a}$$

$$w_i = \frac{\left(N_{obj} - r_i + 1\right)}{\sum_{k=1}^{N_{obj}} \left(N_{obj} - r_k + 1\right)} \tag{7.3b}$$

where r_i = rank of the ith objective O_i. For example, suppose that there are four objectives for MOLCO: O_1, O_2, O_3 and O_4. If O_4 is the least significant, O_1 is the most significant objective among the four, and O_2 is more important than O_3, then r_1, r_2, r_3 and r_4 are 1, 2, 3, and 4, respectively. According to Eq. (7.3a), the weight factor w_i of each objective becomes 0.48, 0.24, 0.16 and 0.12, respectively. By applying Eq. (7.3b) instead of Eq. (7.3a), the weight factors $\mathbf{w} = \{0.4, 0.3, 0.2, 0.1\}$ are obtained.

In the rating method, the value of N_{obj} is assigned to the most important objective. The value decreased by one is assigned to the next important objective. The least important objective has a value of 1. The weight factor w_i is estimated as (Yoon and Hwang 1995)

$$w_i = \frac{v_i}{\sum_{k=1}^{N_{obj}} v_k} \tag{7.4}$$

where v_i is the integer value of importance for the ith objective. For example, assuming that the integer values of the objectives O_1, O_2, O_3 and O_4 are assigned as $\{4, 3, 2, 1\}$ in the order of importance (i.e., O_1 is the most important, whereas O_4 is the least important), the weight factors $w = \{0.4, 0.3, 0.2, 0.1\}$ are obtained.

The paired comparison method is based on a comparison matrix $\mathbf{M}$. The normalized eigenvector associated with the largest eigenvalue of $\mathbf{M}$ becomes the weight factors $\mathbf{w}$. The comparison matrix $\mathbf{M}$ is defined as (Saaty 1977, 2003)

$$\mathbf{M} = \begin{bmatrix} 1 & \alpha_{12} & \cdots & a_{1N_{obj}} \\ \alpha_{21} & 1 & \cdots & a_{2N_{obj}} \\ \vdots & \vdots & 1 & \vdots \\ a_{N_{obj}1} & a_{N_{obj}2} & \cdots & 1 \end{bmatrix} \tag{7.5}$$

where α_{ij} is the relative importance intensity comparing the *i*th objective O_i with the *j*th objective O_j, and is equal to $1/\alpha_{ji}$. When O_i is more important than O_j, the relative importance intensity α_{ij} can range from 2 to 9. The relative importance intensity α_{ij} equal to one indicates that O_i is as important as O_j, and an α_{ij} value of 9 denotes that O_i is absolutely more important than O_j. Suppose that the four objectives functions O_1, O_2, O_3 and O_4 are used for MOLCO, and that the comparison matrix **M** is formulated as

$$\mathbf{M} = \begin{bmatrix} 1 & 9 & 7 & 5 \\ 1/9 & 1 & 5 & 3 \\ 1/7 & 1/5 & 1 & 3 \\ 1/5 & 1/3 & 1/3 & 1 \end{bmatrix} \tag{7.6}$$

As shown in Eq. (7.6), O_1 is treated as the most important, since α_{12}, α_{13} and α_{14} are 9, 7 and 5, respectively. O_2 is more important than O_3, and the associated importance intensities α_{23} and α_{32} are 5 and 0.2, respectively. The largest eigenvalue of **M** is 4.73, and the associated eigenvector is {–0.96, –0.26, –0.12, –0.08}. As a result, the normalized eigenvector becomes the weight factors **w** = {0.68, 0.18, 0.08, 0.06}.

7.2.2 Application to Optimum Inspection Planning

The optimum inspection planning based on the decision making before solving MOLCO is illustrated with a fatigue-sensitive detail of the I-64 Bridge presented in Sub-section 5.2.4. Six objectives are used to establish the optimum inspection plans: $O_{I,1}$ = maximizing the lifetime probability of fatigue crack damage detection P_{det} (see Eq. (3.12)); $O_{I,2}$ = minimizing the expected fatigue crack damage detection delay $E(t_{del})$ (see Eq. (3.19)); $O_{I,3}$ = minimizing the fatigue crack damage detection time-based probability of failure P_f (see Eq. (3.25)); $O_{I,4}$ = minimizing the expected maintenance delay $E(t_{mdl})$ (see Eq. (5.1)); $O_{I,5}$ = maximizing the expected extended service life $E(t_{life})$ (see Eq. (5.3)); and $O_{I,6}$ = minimizing the expected life-cycle cost $E(C_{lcc})$ (see Eq. (5.5)). These six objectives (i.e., $O_{I,1}$ to $O_{I,6}$) are converted into a single objective O_I using the weight sum method as indicated in Eq. (7.1). The formulation of the converted single objective optimization for inspection planning is

$$\text{Find } \mathbf{t_{insp}} = \{t_{insp,1}, t_{insp,2}, \ldots, t_{insp,Ninsp}\} \tag{7.7a}$$

$$\text{for } \mathrm{O_I} = \sum_{i=1}^{6} w_i \mathrm{O_{I,i}}$$

$$= \text{minimizing} \begin{Bmatrix} -\mathrm{w}_1 \cdot P_{\det} + \mathrm{w}_2 \cdot E(t_{del}) + \mathrm{w}_3 \cdot P_f \\ + \mathrm{w}_4 \cdot E(t_{mdl}) - \mathrm{w}_5 \cdot E(t_{life}) + \mathrm{w}_6 \cdot E(C_{lcc}) \end{Bmatrix} \tag{7.7b}$$

$$\text{such that } t_{insp,i} - t_{insp,i-1} \geq 1 \text{ year and } t_{insp,i} \leq 20 \text{ years} \tag{7.7c}$$

$$\text{given } N_{insp} = 1, 2, \text{ or } 3, \text{ and UL inspection method} \tag{7.7d}$$

where $\mathbf{t_{insp}}$ is the inspection times $t_{insp,i}$ (years). The time interval between inspections (i.e., $t_{insp,i} - t_{insp,i-1}$) should be larger than 1 year, and the inspection time $t_{insp,i}$ should be less than 20 years (see Eq. (7.7c)). The number of inspections N_{insp} and the type of the inspection (i.e., UL inspection) are given as indicated in Eq. (7.7d). These converted single-objective optimization problems are solved using the constrained nonlinear minimization algorithm in MATLAB® version 2016b (MathWorks 2016), and the genetic algorithm (GA) in this software is used to check whether the solution is global. It should be noted that the objectives $\mathrm{O_{I,1}}$ (i.e., maximizing P_{det}) and $\mathrm{O_{I,5}}$ (i.e., maximizing $E(t_{life})$) are equivalent to minimizing ($-P_{det}$) and ($-E(t_{life})$) in Eq. (7.7b), respectively.

The weight factors of the six objectives $\mathrm{O_{I,1}}$ to $\mathrm{O_{I,6}}$ are provided in Table 7.1 and Figure 7.2. The weight factors w_i for Cases R-A, R-B, and R-C are based on the ranking method (see Eq. (7.3a)), whereas those for Cases V-A, V-B, and V-C are associated with the rating method (see Eq. (7.4)). The objective $\mathrm{O_{I,1}}$ is treated as the most important among the six objectives (i.e., $r_1 = 1$ and $v_1 = 6$), whereas the objective $\mathrm{O_{I,6}}$ is considered the least important (i.e., $r_6 = 6$ and $v_6 = 1$) in Cases R-A and V-A as shown in Table 7.1 and Figure 7.2. Conversely, Cases R-B and V-B indicate that the objective $\mathrm{O_{I,1}}$ is the least significant (i.e., $r_1 = 6$ and $v_1 = 1$), and that $\mathrm{O_{I,6}}$ is the most significant among the six objectives (i.e., $r_6 = 1$ and $v_6 = 6$). The objective $\mathrm{O_{I,3}}$ has the highest rank of

Table 7.1 Weight factors of objectives for optimum inspection planning resulted from ranking and rating methods.

Ranking method	**Objectives**	**Case R-A**		**Case R-B**		**Case R-C**	
		Rank of objective r_i	**Weight factor w_i**	**Rank of objective r_i**	**Weight factor w_i**	**Rank of objective r_i**	**Weight factor w_i**
	$O_{I,1}$	1	0.41	6	0.07	5	0.08
	$O_{I,2}$	2	0.20	5	0.08	3	0.14
	$O_{I,3}$	3	0.14	4	0.10	1	0.41
	$O_{I,4}$	4	0.10	3	0.14	2	0.2
	$O_{I,5}$	5	0.08	2	0.20	4	0.10
	$O_{I,6}$	6	0.07	1	0.41	6	0.07
Rating method	**Objectives**	**Case V-A**		**Case V-B**		**Case V-C**	
		Value of importance for objective v_i	**Weight factor w_i**	**Value of importance for objective v_i**	**Weight factor w_i**	**Value of importance for objective v_i**	**Weight factor w_i**
	$O_{I,1}$	6	0.29	1	0.05	2	0.10
	$O_{I,2}$	5	0.24	2	0.10	4	0.19
	$O_{I,3}$	4	0.19	3	0.14	6	0.29
	$O_{I,4}$	3	0.14	4	0.19	5	0.24
	$O_{I,5}$	2	0.10	5	0.24	3	0.14
	$O_{I,6}$	1	0.05	6	0.29	1	0.05

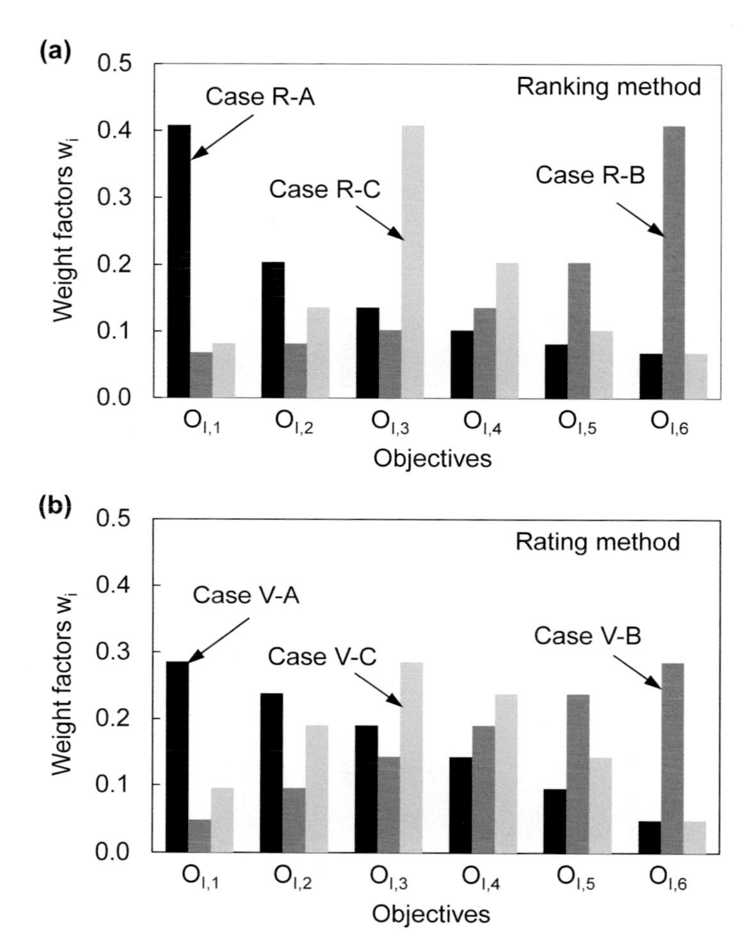

Figure 7.2 Weight factors for optimum inspection planning: (a) ranking method; (b) rating method.

$r_3 = 1$ and the largest importance value of $v_3 = 6$ among the six objectives for both Case R-C and Case V-C, as indicated in Table 7.1 and Figure 7.2.

When the weight factors w_i in Table 7.1 are applied for the converted single objective optimization in Eq. (7.7), the objective values and the UL inspection times are presented in Table 7.2 and Figure 7.3. The application of weight factors for Case R-A ($w_1 = 0.41$, $w_2 = 0.20$, $w_3 = 0.14$, $w_4 = 0.10$,

Table 7.2 Inspection times and the associated objective values for optimum inspection planning based on weight factors in Table 7.1.

Cases	Optimum solutions	UL inspection times (years)			Objective values					
		$t_{insp,1}$	$t_{insp,2}$	$t_{insp,3}$	P_{det}	$E(t_{del})$ (years)	P_f	$E(t_{mdl})$ (years)	$E(t_{life})$ (years)	$E(C_{lcc})$ (USD)
R-A	$I_{R\text{-}A,1}$	14.24	-	-	0.82	10.48	0.16	12.15	36.43	180,345
	$I_{R\text{-}A,2}$	12.16	16.81	-	0.91	7.69	0.07	12.21	37.21	170,546
	$I_{R\text{-}A,3}$	11.26	14.13	18.18	0.94	6.48	0.04	12.79	37.08	173,417
R-B	$I_{R\text{-}B,1}$	14.60	-	-	0.82	10.58	0.17	12.11	36.81	184,759
	$I_{R\text{-}B,2}$	12.83	16.59	-	0.91	8.02	0.08	11.96	37.33	168,122
	$I_{R\text{-}B,3}$	12.38	14.92	17.17	0.94	7.18	0.05	12.17	37.14	168,208
R-C	$I_{R\text{-}C,1}$	14.45	-	-	0.82	10.53	0.17	12.12	36.65	182,952
	$I_{R\text{-}C,2}$	12.74	17.67	-	0.91	7.97	0.08	11.93	37.77	176,464
	$I_{R\text{-}C,3}$	11.96	15.07	19.00	0.94	6.90	0.05	12.30	37.71	176,746
V-A	$I_{V\text{-}A,1}$	14.45	-	-	0.82	10.53	0.17	12.12	36.64	182,864
	$I_{V\text{-}A,2}$	12.38	17.26	-	0.91	7.79	0.07	12.09	37.49	173,480
	$I_{V\text{-}A,3}$	11.26	14.43	19.00	0.94	6.50	0.04	12.75	37.53	179,164
V-B	$I_{V\text{-}B,1}$	14.91	-	-	0.81	10.68	0.17	12.11	37.10	189,614
	$I_{V\text{-}B,2}$	13.21	17.32	-	0.91	8.22	0.08	11.81	37.78	173,314
	$I_{V\text{-}B,3}$	12.53	15.37	17.93	0.94	7.27	0.06	12.06	37.51	171,461
V-C	$I_{V\text{-}C,1}$	14.56	-	-	0.82	10.56	0.17	12.11	36.76	184,337
	$I_{V\text{-}C,2}$	12.77	18.24	-	0.91	8.00	0.08	11.90	38.03	182,827
	$I_{V\text{-}C,3}$	12.06	15.62	19.96	0.94	6.99	0.05	12.21	38.27	184,087

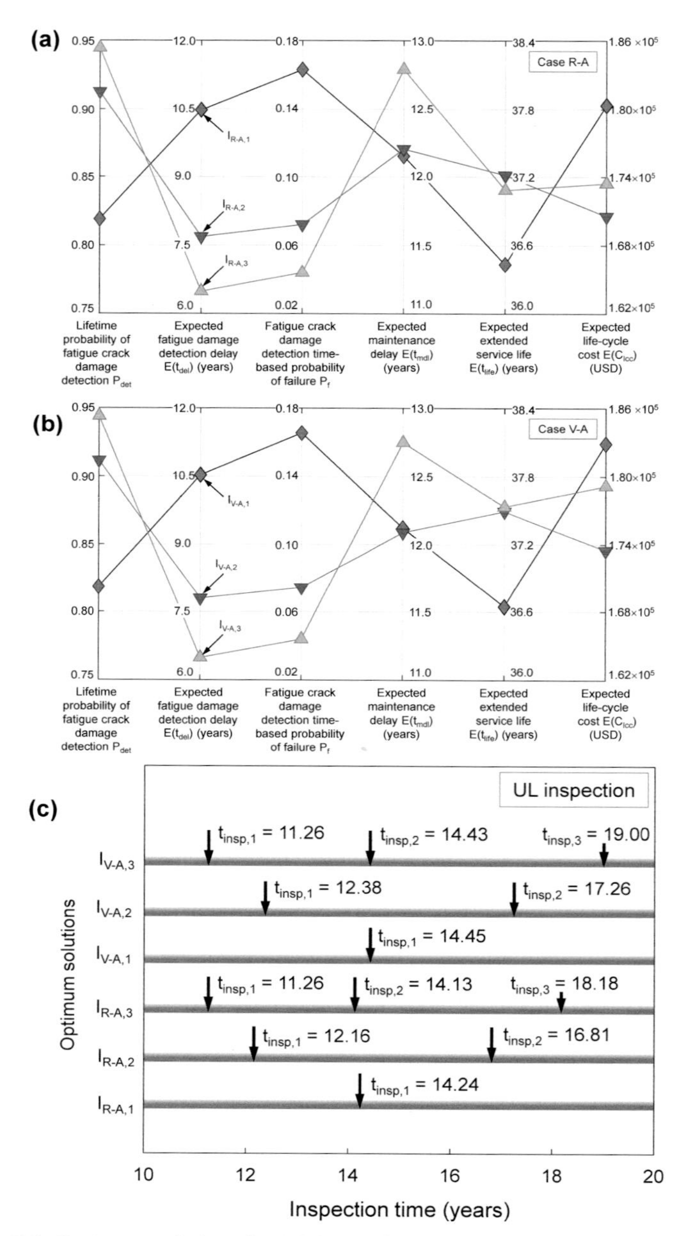

Figure 7.3 Optimum solutions for UL inspection planning with N_{insp} = 1, 2 and 3: (a) case R-A; (b) case V-A; (c) inspection plans.

w_5 = 0.08, and w_6 = 0.07) leads to the optimum solutions $I_{R\text{-}A,1}$, $I_{R\text{-}A,2}$ and $I_{R\text{-}A,3}$, which are based on the given number of inspections N_{insp} = 1, 2, and 3, respectively. The objective values of the solutions $I_{R\text{-}A,1}$, $I_{R\text{-}A,2}$ and $I_{R\text{-}A,3}$ are illustrated in the parallel coordinate system as shown in Figure 7.3(a). The associated UL inspection plans are shown in Figure 7.3(c). The solution $I_{R\text{-}A,1}$ requires the one time UL inspection at 14.24 years, and results in P_{det} = 0.82, $E(t_{del})$ = 10.48 years, P_f = 0.16, $E(t_{mdl})$ = 12.15 years, $E(t_{life})$ = 36.43 years and $E(C_{lcc})$ = \$1.80 × 10^5 (see Table 7.2 and Figures 7.3(a) and 7.3(c)). The three-time UL inspections at 11.26, 14.13, and 18.18 years are represented by the solution $I_{R\text{-}A,3}$. If the weight factors for Case V-A are used instead of those for Case R-A, the three-time UL inspections have to be performed at 11.26, 14.43, and 19.00 years, and, as a result, P_{det} = 0.94, $E(t_{del})$ = 6.50 years, P_f = 0.04, $E(t_{mdl})$ = 12.75 years, $E(t_{life})$ = 37.53 years and $E(C_{lcc})$ = \$1.79 × 10^5 can be obtained (see solution $I_{V\text{-}A,3}$ in Table 7.2 and Figures 7.3(b) and 7.3(c)). Figure 7.4(a) illustrates the objective values for the solutions $I_{R\text{-}A,2}$, $I_{R\text{-}B,2}$ and $I_{R\text{-}C,2}$, which are associated with N_{insp} = 2 and w_i for Cases R-A, R-B, and R-C, in the parallel coordinate system. The optimum solutions $I_{V\text{-}A,2}$, $I_{V\text{-}B,2}$ and $I_{V\text{-}C,2}$ are based on N_{insp} = 2 and w_i for Cases V-A, V-B, and V-C, and the corresponding objective values are presented in Figure 7.4(b). From Table 7.2 and Figures 7.3 and 7.4, it can be seen that the optimum inspection plans depend on the weight factors of the objectives.

7.2.3 Application to Optimum Monitoring Planning

The decision making before solving MOLCO is applied for optimum monitoring planning of the ship hull structure under fatigue presented in Sub-section 3.4.4. The detailed information on the probabilistic fatigue crack propagation over time and the life-cycle cost estimation for this ship structure can be found in Table 3.1 and Table 5.3. The optimum monitoring planning is based on the following five objectives: $O_{M,1}$ = minimizing the expected fatigue crack damage detection delay $E(t_{del})$ (see Eq. (3.20)); $O_{M,2}$ = minimizing the fatigue crack damage detection time-based probability of failure P_f (see Eq. (3.25)); $O_{M,3}$ = minimizing the expected maintenance delay $E(t_{mdl})$ (see Eq. (5.8)); $O_{M,4}$ = maximizing the expected extended service life

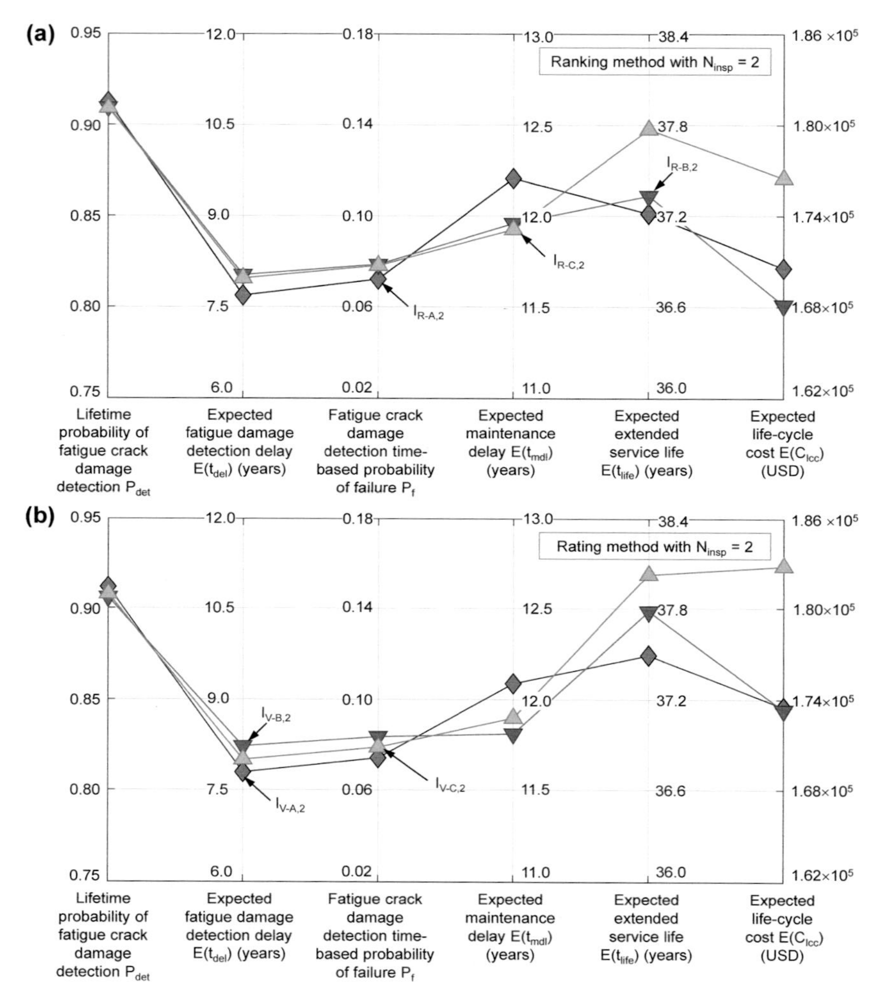

Figure 7.4 Optimum solutions for UL inspection planning with $N_{insp} = 2$ in the parallel coordinate system: (a) ranking method; (b) rating method.

$E(t_{life})$ (see Eq. (5.10)); and $O_{M,5}$ = minimizing the expected life-cycle cost $E(C_{lcc})$ (see Eq. (5.11)).

By converting the five objectives (i.e., $O_{M,1}$ to $O_{M,5}$) into a single objective O_M using the weight sum method, the optimization for monitoring planning is formulated as

$$\text{Find } \mathbf{t}_{\mathbf{ms}} = \{t_{ms,1}, t_{ms,2}, \ldots, t_{ms,Nmon}\} \tag{7.8a}$$

$$\text{for } \mathrm{O_M} = \sum_{i=1}^{5} w_i \mathrm{O_{M,i}}$$

$$= \text{minimizing} \begin{Bmatrix} \mathrm{w}_1 \cdot E\left(t_{del}\right) + \mathrm{w}_2 \cdot P_f \\ + \mathrm{w}_3 \cdot E\left(t_{mdl}\right) + \mathrm{w}_4 \cdot \left[-E\left(t_{life}\right)\right] + \mathrm{w}_5 \cdot E\left(C_{lcc}\right) \end{Bmatrix} \tag{7.8b}$$

$$\text{such that } 1 \text{ year} \le t_{ms,i} - (t_{ms,i-1} + t_{md}) \le 20 \text{ years} \tag{7.8c}$$

$$\text{given } N_{mon} = 1, 2 \text{ or } 3, \text{ and } t_{md} = 0.3 \text{ years} \tag{7.8d}$$

The design variables are the monitoring starting times $t_{ms,i}$ (years). The non-monitoring time interval (i.e., $t_{ms,i} - (t_{ms,i-1} + t_{md})$) should be larger than 1 year and less than 20 years as indicated in Eq. (7.8c). The number of monitorings N_{mon} = 1, 2 or 3 and the monitoring duration t_{md} = 0.3 year are given (see Eq. (7.8d)). The constrained nonlinear minimization algorithm in MATLAB version 2016b (MathWorks 2016) is used to solve the single-objective optimization in Eq. (7.8).

The weight factors $\mathbf{w} = \{w_1, w_2, w_3, w_4, w_5\}$ in Eq. (7.8b) are determined using the paired comparison method as shown in Table 7.3. For Case P-A, the objective $\mathrm{O_{M,1}}$ is absolutely the most important among the five objectives; $\mathrm{O_{M,2}}$ is more important than $\mathrm{O_{M,3}}$, $\mathrm{O_{M,4}}$, and $\mathrm{O_{M,5}}$; and $\mathrm{O_{M,5}}$ is the least important. Accordingly, the relative importance intensities α_{12}, α_{13}, α_{14} and α_{15} in the comparison matrix **M** become 9, 7, 5, and 3, respectively. By computing the normalized eigenvector associated with the largest eigenvalue of **M**, the weight factors **w** can be obtained as {0.53, 0.26, 0.11, 0.06, 0.04} as shown in Table 7.3. The Case P-B indicates that the objective $\mathrm{O_{M,5}}$ is the most important among the five objectives, and that $\mathrm{O_{M,1}}$ is the least significant objective. The objective $\mathrm{O_{M,3}}$ is treated as the most important in the Case P-C. Figure 7.5 compares the weight factors **w** for Cases P-A, P-B and P-C.

The objective values and the monitoring plans associated with the optimum solutions for Cases P-A, P-B, and P-C are provided in Table 7.4. Figure 7.6(a) shows the objective values of the solutions $\mathrm{M_{P\text{-}A,2}}$, $\mathrm{M_{P\text{-}B,2}}$ and $\mathrm{M_{P\text{-}C\text{-}2}}$ in the parallel coordinate system. The monitoring plans of these

Table 7.3 Weight factors of objectives for optimum monitoring planning resulted from paired comparison method.

Case P-A	Comparison matrix	$\mathbf{M} = \begin{bmatrix} 1 & 9 & 7 & 5 & 3 \\ 1/9 & 1 & 9 & 7 & 5 \\ 1/7 & 1/9 & 1 & 7 & 5 \\ 1/5 & 1/7 & 1/7 & 1 & 7 \\ 1/3 & 1/5 & 1/5 & 1/7 & 1 \end{bmatrix}$
	Weight factor	$\mathbf{w} = \{0.53, 0.26, 0.11, 0.06, 0.04\}$
Case P-B	Comparison matrix	$\mathbf{M} = \begin{bmatrix} 1 & 1/7 & 1/7 & 1/9 & 1/9 \\ 7 & 1 & 1/5 & 1/7 & 1/7 \\ 7 & 5 & 1 & 1/2 & 1/2 \\ 9 & 7 & 5 & 1 & 1/3 \\ 9 & 7 & 5 & 3 & 1 \end{bmatrix}$
	Weight factor	$\mathbf{w} = \{0.02, 0.06, 0.13, 0.31, 0.47\}$
Case P-C	Comparison matrix	$\mathbf{M} = \begin{bmatrix} 1 & 1/7 & 1/5 & 1/3 & 1/9 \\ 7 & 1 & 1/9 & 1/5 & 1/3 \\ 5 & 9 & 1 & 7 & 3 \\ 3 & 5 & 1/7 & 1 & 5 \\ 9 & 3 & 1/3 & 1/5 & 1 \end{bmatrix}$
	Weight factor	$\mathbf{w} = \{0.04, 0.07, 0.52, 0.23, 0.14\}$

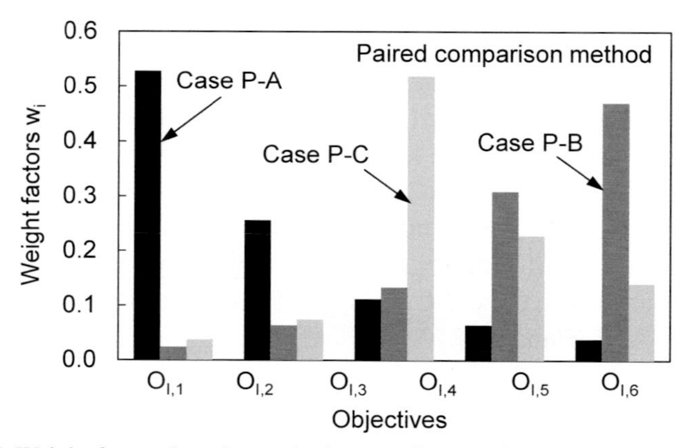

Figure 7.5 Weight factors based on paired comparison method for optimum monitoring planning.

Table 7.4 Monitoring starting times and the associated objective values for optimum monitoring planning based on weight factors in Table 7.3.

Cases	Optimum solutions	Optimum monitoring starting times (years)			Objective values				
		$t_{ms,1}$	$t_{ms,2}$	$t_{ms,3}$	$E(t_{del})$ (years)	P_f	$E(t_{mdl})$ (years)	$E(t_{life})$ (years)	$E(C_{lcc})$ (USD)
P-A	$M_{P\text{-}A,1}$	7.18	-	-	5.27	0.13	9.12	19.21	698900
	$M_{P\text{-}A,2}$	5.21	10.80	-	2.96	0.05	7.53	24.43	316931
	$M_{P\text{-}A,3}$	3.75	7.76	14.38	2.00	0.03	6.65	27.68	248221
P-B	$M_{P\text{-}B,1}$	9.70	-	-	6.46	0.22	9.01	21.05	322460
	$M_{P\text{-}B,2}$	7.10	14.49	-	3.83	0.11	7.27	27.90	247730
	$M_{P\text{-}B,3}$	5.73	10.99	15.77	2.71	0.07	6.30	27.85	223618
P-C	$M_{P\text{-}B,1}$	9.06	-	-	6.05	0.19	8.91	20.71	401743
	$M_{P\text{-}B,2}$	7.11	13.54	-	3.81	0.11	7.14	26.40	235634
	$M_{P\text{-}B,3}$	6.02	10.46	15.23	2.80	0.08	6.13	27.64	240151

solutions are illustrated in Figure 7.6(b). When the weight factors **w** for Case P-A and the number of monitorings N_{mon} = 2 are used to solve the optimization problem defined in Eq. (7.8), the optimum monitoring starting times obtained are $t_{ms,1}$ = 5.21 years and $t_{ms,2}$ = 10.80 years, which are represented by the optimum solution $M_{P\text{-}A,2}$ in Table 7.4

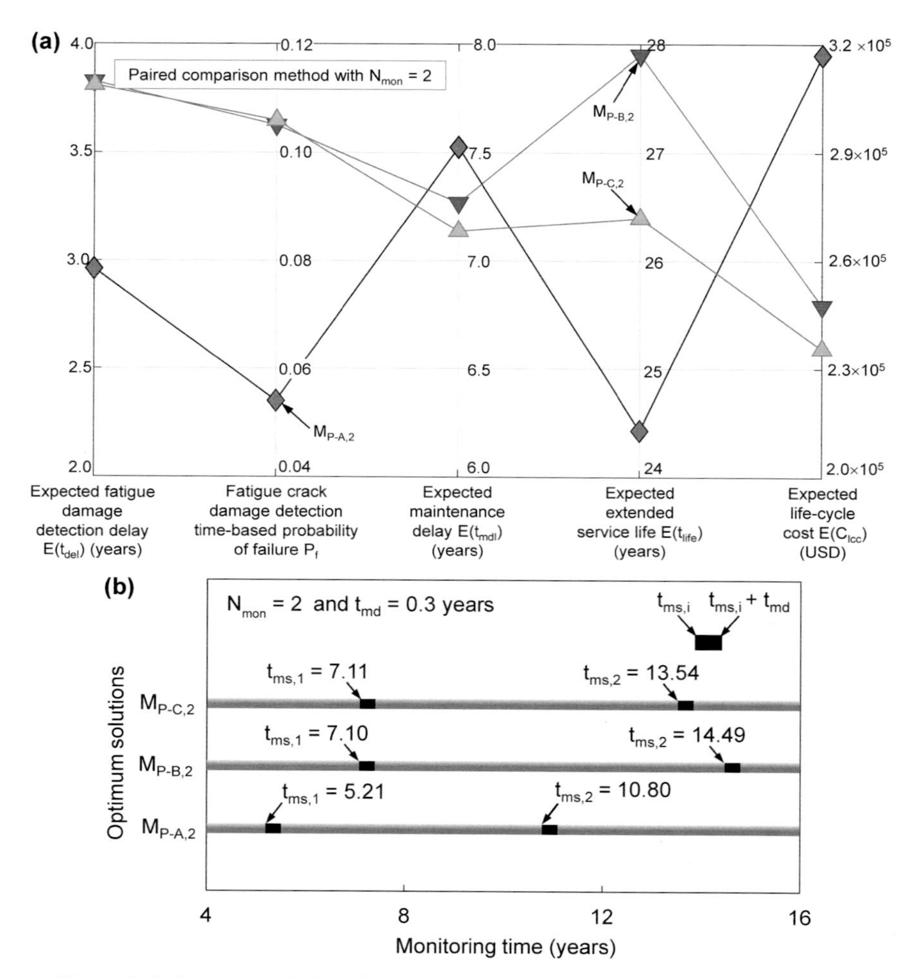

Figure 7.6 Optimum solutions for monitoring planning based on paired comparison method with N_{mon} = 2 and t_{md} = 0.3 year: (a) objective values in parallel coordinate system; (b) monitoring plans for $M_{P\text{-}A,2}$, $M_{P\text{-}B,2}$ and $M_{P\text{-}C,2}$.

and Figure 7.6. The associated objective values become $E(t_{del}) = 2.96$ years, $P_f = 0.05$, $E(t_{mdl}) = 7.53$ years, $E(t_{life}) = 24.43$ years and $E(C_{lcc}) = \$3.17 \times 10^5$. If the weight factors **w** for Case P-B are used instead of those for Case P-A, the monitoring starting times $t_{ms,1} = 7.10$ years and $t_{ms,2} = 14.49$ years can be computed (see the solution $M_{P\text{-}B,2}$ in Table 7.4 and Figure 7.6). Furthermore, the optimum solution $M_{P\text{-}C,2}$ based on the weight factors **w** for Case P-C indicates the monitoring starting times $t_{ms,1} = 7.11$ years and $t_{ms,2} = 13.54$ years.

7.3 Decision Making after Multi-Objective Life-Cycle Optimization

Decision making after solving MOLCO is performed to identify the essential objectives, determine the weight factors of the essential objectives, and select the best Pareto solution for the optimum inspection and monitoring plan as shown in Figure 7.1. By using the objective reduction approach, the essential objectives and redundant objectives can be identified. To determine the weight factors of the essential objectives, the objective method for weight factor determination (e.g., SD, CRITIC, and CCSD methods) is used. MADM results in the best Pareto optimal solution, which represents the optimum inspection and monitoring application time.

7.3.1 Essential and Redundant Objectives

An increase in the number of objectives can lead to high computational cost, low efficiency in searching the Pareto front and difficulties in visualization and decision making. These drawbacks can be addressed by using the objective reduction approach. This approach leads to identifying the essential and redundant objectives among the initial objective set. The Pareto front of the multi-objective optimization is affected only by the essential objectives. The redundant objectives have no effect on the Pareto front. The Pareto front considering the essential objectives will be the same as the Pareto front for the initial objective set. In this study, the objective reduction approach developed by Brockhoff and Zitzler (2006, 2007, 2009), which is based on the dominance relation among the objective values, is used.

The Pareto optimal solution set can be expressed using the dominance relation among the objective values. Suppose that there is an initial objective set $\boldsymbol{\Omega}_{\mathrm{I}}$ consisting of N_{obj} objectives (i.e., $\boldsymbol{\Omega}_{\mathrm{I}} = \{f_1, f_2, \ldots, f_{N_{obj}}\}$) to be minimized. The Pareto optimal solution set Φ_{sl} and the Pareto front Φ_{ft} are defined as

$$\Phi_{sl} := \{\mathbf{x} \in \mathbf{X} | \nexists\ \mathbf{y} \in \mathbf{X} : \mathbf{y} \prec \mathbf{x}\} \tag{7.9a}$$

$$\Phi_{ft} := \{\mathbf{z} = [f_1(\mathbf{x}), f_2(\mathbf{x}), \ldots, f_{N_{obj}}(\mathbf{x})] | \mathbf{x} \in \Phi_{sl}\} \tag{7.9b}$$

where $\mathbf{x}$ = Pareto optimal solution; $\mathbf{X}$ = design space. $\mathbf{y} \prec \mathbf{x}$ denotes that $\mathbf{y}$ dominates $\mathbf{x}$. If and only if $f_i(\mathbf{y}) \leq f_i(\mathbf{x})$ for all objective functions of the initial objective set $\boldsymbol{\Omega}_{\mathrm{I}}$, and $f_i(\mathbf{y}) < f_i(\mathbf{x})$ for at least one objective function of $\boldsymbol{\Omega}_{\mathrm{I}}$, then a solution $\mathbf{y}$ is said to dominate $\mathbf{x}$. Equation (7.9a) indicates that $\mathbf{x} \in \mathbf{X}$ is the Pareto optimal solution if there is no solution $\mathbf{y} \in \mathbf{X}$ to dominate $\mathbf{x}$. The set of the objective values for $\mathbf{x} \in \Phi_{\mathrm{sl}}$ is the Pareto front Φ_{ft}, as indicated in Eq. (7.9b).

The degree of conflict Δ between the initial objective set $\boldsymbol{\Omega}_{\mathrm{I}}$ and the reduced objective set $\boldsymbol{\Omega}_{\mathrm{R}} \subseteq \boldsymbol{\Omega}_{\mathrm{I}}$ is estimated in the dominance relation-based objective reduction approach, which is expressed as the maximum difference between the Pareto optimal solutions of $\boldsymbol{\Omega}_{\mathrm{R}}$ and those of $\boldsymbol{\Omega}_{\mathrm{I}}$. The degree of conflict Δ should be normalized, when the objective values are expressed with different units and scales. When $\boldsymbol{\Omega}_{\mathrm{I}}$ completely conflicts with $\boldsymbol{\Omega}_{\mathrm{R}}$, the normalized Δ_{norm} between $\boldsymbol{\Omega}_{\mathrm{I}}$ and $\boldsymbol{\Omega}_{\mathrm{R}}$ is 1.0. $\boldsymbol{\Omega}_{\mathrm{R}}$ associated with $\Delta_{\mathrm{norm}} = 0$ (i.e., non-conflicting) results in the Pareto front being the same as the Pareto front Φ_{ft} of $\boldsymbol{\Omega}_{\mathrm{I}}$. The essential objective set $\boldsymbol{\Omega}^{*}_{\mathrm{R}}$ is defined as the reduced objective set $\boldsymbol{\Omega}_{\mathrm{R}}$ with the smallest number of objectives that can induce the same Pareto front Φ_{ft} of $\boldsymbol{\Omega}_{\mathrm{I}}$ (i.e., $\Delta_{\mathrm{norm}} = 0$). The non-essential objectives in $\boldsymbol{\Omega}_{\mathrm{I}}$ are treated as redundant. It should be noted that the dominance relation-based objective reduction approach requires the Pareto optimal solutions obtained from the multi-objective optimization considering $\boldsymbol{\Omega}_{\mathrm{I}}$. The detailed theoretical background and the computational procedure of the aforementioned approach can be found in Brockhoff and Zitzler (2006, 2007, 2009).

7.3.2 Weights of Essential Objectives

By using the objective weight determination methods such as the SD, CRITIC and CCSD methods, the weights of the essential objectives are determined. The redundant objectives are ignored in MADM. Therefore, the weights of the redundant objectives are zero, since the redundant objectives have no contribution to the Pareto optimal front. The objective weight determination methods use the decision information including the correlation among the objective values of the Pareto optimal solutions Φ_{sl} and the SD of the objective values of Φ_{sl}.

Under the premise that a larger SD of the objective values of Φ_{sl} has a more significant impact on decision making, and therefore, the associated objective needs to have more weight in MADM, the SD method determines the weight w_i of the ith essential objective as (Deng et al. 2000)

$$w_i = \frac{\sigma_i}{\sum_{j=1}^{N^*_{obj}} \sigma_j} \tag{7.10}$$

where σ_i = SD of the ith essential objective values z_i of Φ_{sl}; N^*_{obj} = number of essential objectives. The CRITIC method is based on the assumption that the objective less correlated with other objectives results in a more significant impact on decision making as well as based on the assumption for the SD method. Using the CRITIC method, the weight w_i of the ith essential objective is estimated as (Diakoulaki et al. 1995)

$$w_i = \frac{\sigma_i \sum_{j=1}^{N^*_{obj}} \left(1 - \chi_{ij}\right)}{\sum_{j=1}^{N^*_{obj}} \left(\sigma_j \sum_{k=1}^{N^*_{obj}} \left(1 - \chi_{jk}\right) \right)} \tag{7.11}$$

where χ_{ij} is the coefficient of correlation between the *i*th essential objective values z_i and the *j*th essential objective values z_j of Φ_{sl}. Furthermore, in the CCSD method, the weight w_i is computed as (Wang and Luo 2010)

$$w_i = \frac{\sigma_i \sqrt{1-\chi_i}}{\sum_{j=1}^{N^*_{obj}} \left(\sigma_j \sqrt{1-\chi_j}\right)} \tag{7.12}$$

where χ_i is the coefficient of correlation between z_i and V_{ij}. V_{ij} is defined as

$$V_{ij} = \sum_{k=1,k\neq i}^{N^*_{obj}} w_k z_{ki} \tag{7.13}$$

where z_{ki} is the *k*th normalized value of the *i*th essential objective values z_i of Φ_{sl}. In order to estimate χ_i in Eq. (7.12), V_{ij} should be estimated with the weight w_i (see Eq. (7.13)). For this reason, the optimization process is applied to compute w_i as follows:

Find $\boldsymbol{w} = \{w_1, \ldots, w_j, \ldots, w_{Nobj}\}$ (7.14a)

for minimizing $\sum_{i=1}^{N^*_{obj}} \left(w_i - \frac{\sigma_i \sqrt{1-\chi_i}}{\sum_{j=1}^{N^*_{obj}} \left(\sigma_j \cdot \sqrt{1-\chi_j}\right)} \right)^2$ (7.14b)

such that $w_j \geq 0$ and $\sum_{i=1}^{N^*obj} w_i = 1$ (7.14c)

7.3.3 *Multiple Attribute Decision Making*

Using the MADM process, the best optimal solution can be selected from the Pareto set. The MADM methods used in this chapter are the simple additive weighting (SAW) method, the technique for order preference by similarity to ideal solution (TOPSIS) method, and the elimination and choice expressing the reality (ELECTRE) method. These three methods are the most widely used MADM methods (Zanakis et al. 1998; Yeh 2002). In the SAM method, the overall assessment value of the *i*th Pareto optimal solution A_i is estimated as (Yoon and Hwang 1995)

$$A_i = \sum_{k=1}^{N^*_{obj}} w_k f_{ki}^{norm} \quad (7.15)$$

where f_{ki}^{norm} = *k*th normalized objective value of the *i*th Pareto optimal solution. The weight of the *k*th objective w_k is obtained using Eqs. (7.10), (7.11) and (7.12), which are associated with the SD, CRITIC, and CCSD methods, respectively. The Pareto optimal solution with the largest overall assessment value A_i corresponds to the best Pareto optimal solution.

The TOPSIS method is based on the distance between the *i*th Pareto optimal solution and the ideal solution. The overall assessment value of the *i*th Pareto optimal solution A_i is computed as

$$A_i = \frac{d_i^-}{d_i^+ + d_i^-} \quad (7.16)$$

where d_i^+ is the distance between the *i*th Pareto optimal solution and the positive ideal solution, and d_i^- is the distance between the *i*th Pareto optimal solution and the negative ideal solution. Estimation of d_i^+ and d_i^- is based on the weighted normalized objective values. The best Pareto optimal solution has the largest A_i. More information on the TOPSIS method can be found in Hwang et al. (1993).

The ELECTRE method requires performing the aggregation and exploitation procedures. The aggregation procedure uses the concordance and discordance indexes to formulate the pairwise comparisons of the Pareto optimal set, and to find the comprehensive outranking relations by considering three cases: preference, indifference, and incomparability. The exploitation procedure results in the binary outranking relationships among the Pareto optimal solutions. Finally, the overall ranking of the Pareto optimal solutions can be estimated. The detailed algorithm and the associated theoretical background of the ELECTRE method are provided in Roy (1971), Nijkamp and van Delft (1977) and Voogd (1983).

7.3.4 Application to Optimum Inspection Planning

The decision making after solving MOLCO is applied for optimum inspection planning of an existing bridge, which is the I-64 Bridge presented in Sub-sections 5.2.4 and 7.2.2. The decision making after solving MOLCO is based on the initial objective set Ω_I consisting of the six objectives $O_{I,1}$ to $O_{I,6}$. The formulation of the six objective inspection planning can be found in Sub-section 6.2.4. The Pareto optimal solutions provided in Figure 6.7 are used to identify the essential and redundant objectives through the dominance relation-based objective reduction approach.

Table 7.5 indicates the representative reduced objective set Ω_R associated with the normalized degree of conflict Δ_{norm} = 0 and 1. The reduced objective sets $\Omega_R = \{O_{I,1}, O_{I,2}\}$, $\{O_{I,1}, O_{I,3}\}$, $\{O_{I,2}, O_{I,6}\}$, $\{O_{I,1}, O_{I,2}, O_{I,3}\}$, $\{O_{I,2}, O_{I,3}, O_{I,6}\}$ and $\{O_{I,1}, O_{I,2}, O_{I,3}, O_{I,6}\}$ lead to Δ_{norm} = 1 for one, two and three UL inspections. The dimensionality of the Pareto optimal solutions of Ω_I is six, because Ω_I consists of the six objectives $O_{I,1}$ to $O_{I,6}$. As shown in Figure 7.7(a), the Pareto optimal solutions of Ω_I for N_{insp} = 1 are projected onto two dimensions consisting of P_{det} and $E(t_{del})$ for comparison with the Pareto optimal solutions of $\Omega_R = \{O_{I,1}, O_{I,2}\}$. Figure 7.7(b) compares the Pareto optimal solutions of Ω_I and $\Omega_R = \{O_{I,1}, O_{I,3}\}$ for N_{insp} = 1 in the two dimensional coordinate system associated with P_{det} and P_f. Figure 7.8 shows the Pareto solutions of Ω_I and $\Omega_R = \{O_{I,1}, O_{I,3}, O_{I,4}, O_{I,5}\}$ for N_{insp} = 1 in the two 3D coordinate systems (i.e., $\{P_{det}, P_f, E(t_{mdl})\}$ and $\{P_f, E(t_{mdl}), E(t_{life})\}$). It can be seen that the Pareto front of Ω_I is the same as the Pareto front of $\Omega_R = \{O_{I,1}, O_{I,3}, O_{I,4}, O_{I,5}\}$. This is because the normalized degree of conflict Δ_{norm} between Ω_I and $\Omega_R = \{O_{I,1}, O_{I,3}, O_{I,4}, O_{I,5}\}$ is equal to zero as indicated in Table 7.5. Furthermore, Δ_{norm} between Ω_I and $\Omega_R = \{O_{I,2}, O_{I,4}, O_{I,5}, O_{I,6}\}$ is also equal to zero, and the Pareto fronts of these two objective sets are the same as shown in Figure 7.9. The reduced objective sets $\Omega_R = \{O_{I,1}, O_{I,3}, O_{I,4}, O_{I,5}\}$, $\{O_{I,2}, O_{I,3}, O_{I,4}, O_{I,5}\}$ and $\{O_{I,2}, O_{I,4}, O_{I,5}, O_{I,6}\}$ are the essential objective set for N_{insp} = 1 (see Table 7.5). For N_{insp} = 2 and 3, the essential objective set is $\{O_{I,2}, O_{I,4}, O_{I,5}, O_{I,6}\}$, and the objectives $O_{I,1}$ and $O_{I,3}$ are redundant.

The weight factors of the initial objective set Ω_I for N_{insp} = 2 are computed using the SD, CRITIC, and CCSD methods (see Eqs. (7.10), (7.11) and

Table 7.5 Representative reduced objective sets Ω_R associated with $\Delta_{norm} = 0$ and 1.

Normalized degree of conflict Δ_{norm} between Ω_I and Ω_R	**Number of UL inspections $N_{insp} = 1$**	**Number of UL inspections $N_{insp} = 2$**	**Number of UL inspections $N_{insp} = 3$**
$\Delta_{norm} = 1$ (perfect-conflict)	$\{O_{I,1}, O_{I,2}\}$	$\{O_{I,1}, O_{I,2}\}$	$\{O_{I,1}, O_{I,2}\}$
	$\{O_{I,1}, O_{I,3}\}$	$\{O_{I,1}, O_{I,3}\}$	$\{O_{I,1}, O_{I,3}\}$
	$\{O_{I,2}, O_{I,6}\}$	$\{O_{I,2}, O_{I,6}\}$	$\{O_{I,2}, O_{I,6}\}$
	$\{O_{I,1}, O_{I,2}, O_{I,3}\}$	$\{O_{I,1}, O_{I,2}, O_{I,3}\}$	$\{O_{I,1}, O_{I,2}, O_{I,3}\}$
	$\{O_{I,2}, O_{I,3}, O_{I,6}\}$	$\{O_{I,2}, O_{I,3}, O_{I,6}\}$	$\{O_{I,2}, O_{I,3}, O_{I,6}\}$
	$\{O_{I,1}, O_{I,2}, O_{I,3}, O_{I,6}\}$	$\{O_{I,1}, O_{I,2}, O_{I,3}, O_{I,6}\}$	$\{O_{I,1}, O_{I,2}, O_{I,3}, O_{I,6}\}$
$\Delta_{norm} = 0$ (non-conflict)	$\{O_{I,1}, O_{I,3}, O_{I,4}, O_{I,5}\}$ [a]	$\{O_{I,2}, O_{I,4}, O_{I,5}, O_{I,6}\}$ [b]	$\{O_{I,2}, O_{I,4}, O_{I,5}, O_{I,6}\}$ [b]
	$\{O_{I,2}, O_{I,3}, O_{I,4}, O_{I,5}\}$ [a]	$\{O_{I,1}, O_{I,2}, O_{I,4}, O_{I,5}, O_{I,6}\}$	$\{O_{I,1}, O_{I,2}, O_{I,4}, O_{I,5}, O_{I,6}\}$
	$\{O_{I,2}, O_{I,4}, O_{I,5}, O_{I,6}\}$ [b]	$\{O_{I,2}, O_{I,3}, O_{I,4}, O_{I,5}, O_{I,6}\}$	$\{O_{I,2}, O_{I,3}, O_{I,4}, O_{I,5}, O_{I,6}\}$
	$\{O_{I,3}, O_{I,4}, O_{I,5}, O_{I,6}\}$ [a]		
	$\{O_{I,1}, O_{I,2}, O_{I,3}, O_{I,4}, O_{I,5}\}$		
	$\{O_{I,1}, O_{I,3}, O_{I,4}, O_{I,5}, O_{I,6}\}$		
	$\{O_{I,2}, O_{I,3}, O_{I,4}, O_{I,5}, O_{I,6}\}$		

[a] Essential objective set only for $N_{insp} = 1$.
[b] Essential objective set for $N_{insp} = 1$, 2 and 3.

(7.12), respectively) and are provided in Table 7.6. The weight factors w_1 and w_3 are zero, since the objectives $O_{I,1}$ and $O_{I,3}$ for $N_{insp} = 2$ are redundant, and ignored in the MADM process. The well-balanced optimum solutions from the Pareto optimal solutions are determined through the two MADM methods (i.e., SAW and TOPSIS methods). The solutions associated with the first, second and third largest overall assessment values are presented in Table 7.6. The solutions $I_{S\text{-}S,1}$, $I_{R\text{-}S,1}$ and $I_{C\text{-}S,1}$ have the largest overall assessment value based on the SAW method, when the weight factors are determined using the SD, CRITIC, and CCSD methods, respectively. If the

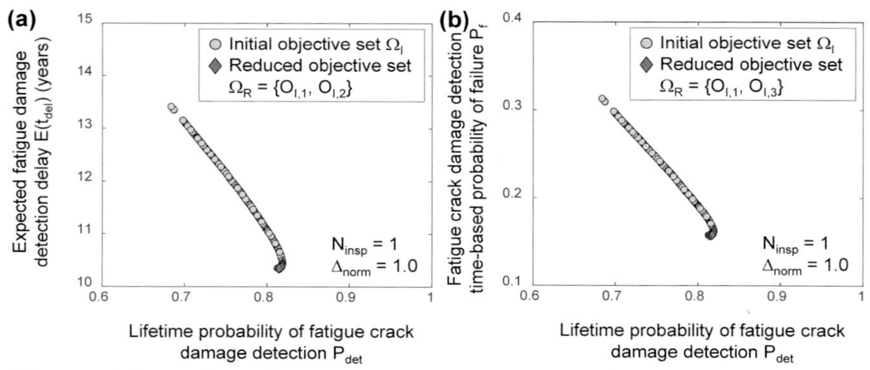

Figure 7.7 Comparison of Pareto optimal solutions for Ω_I and Ω_R associated with N_{insp} = 1: (a) $\Omega_R = \{O_{I,1}, O_{I,2}\}$; (b) $\Omega_R = \{O_{I,1}, O_{I,3}\}$.

Color version at the end of the book

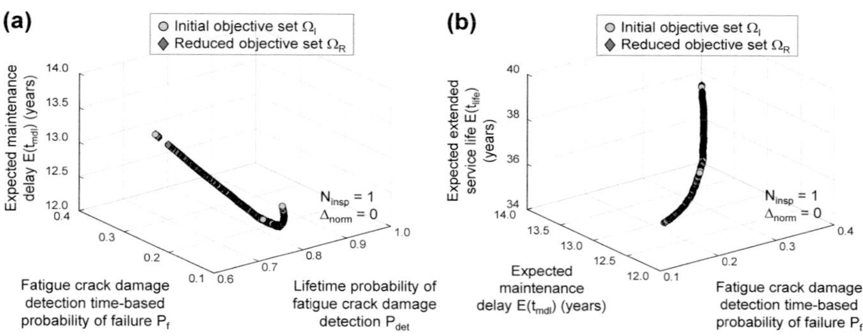

Figure 7.8 Comparison of Pareto optimal solutions for Ω_I and $\Omega_R = \{O_{I,1}, O_{I,3}, O_{I,4}, O_{I,5}\}$ associated with N_{insp} = 1 in 3D coordinate system: (a) $\{P_{det}, P_f, E(t_{mdl})\}$; (b) $\{P_f, E(t_{mdl}), E(t_{life})\}$.

Color version at the end of the book

TOPSIS method is applied instead of the SAW method, the solutions $I_{S\text{-}T,1}$, $I_{R\text{-}T,1}$ and $I_{C\text{-}T,1}$ are associated with the largest overall assessment value. The values of the design variables of each solution (i.e., optimum inspection times $t_{insp,1}$ and $t_{insp,2}$) can be found in Table 7.6. The associated objective values are illustrated in the parallel coordinate system as shown in Figure 7.10. The solution $I_{S\text{-}S,1}$ indicates the optimum inspection times $t_{insp,1}$ = 13.15 years and $t_{insp,2}$ = 19.39 years (see Table 7.6), and the objective

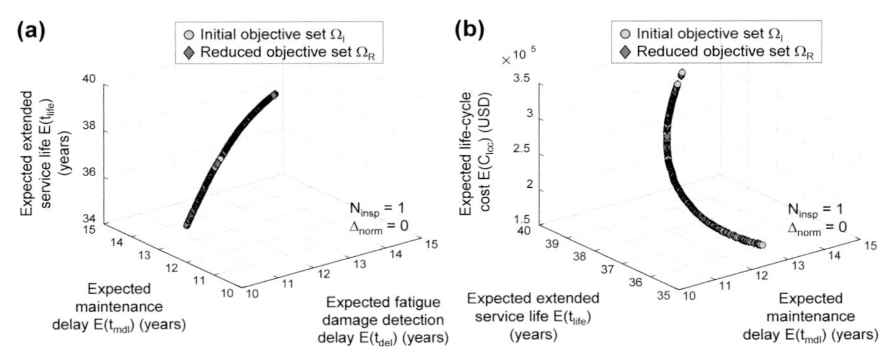

Figure 7.9 Comparison of Pareto optimal solutions for Ω_I and $\Omega_R = \{O_{I,2}, O_{I,4}, O_{I,5}, O_{I,6}\}$ associated with $N_{insp} = 1$ in 3D coordinate system: (a) $\{E(t_{del}), E(t_{mdl}), E(t_{life})\}$; (b) $\{E(t_{mdl}), E(t_{life}), E(C_{lcc})\}$.

Color version at the end of the book

values are $P_{det} = 0.90$, $E(t_{del}) = 8.26$ years, $P_f = 0.09$, $E(t_{mdl}) = 11.77$ years, $E(t_{life}) = 38.65$ years and $E(C_{lcc}) = \$195{,}584$ (see Figure 7.10(a)). The solution $I_{S\text{-}S,1}$ is the same as $I_{R\text{-}S,1}$ and $I_{C\text{-}S,1}$. The solutions $I_{S\text{-}S,2}$, $I_{R\text{-}S,2}$ and $I_{C\text{-}S,2}$ are also identical each other. Furthermore, the solution $I_{S\text{-}T,1}$ in Table 7.6 is selected as the best optimal solution, when the weight factors are determined using the SD method, and the overall assessment value for MADM is estimated by the TOPSIS method. The associated UL inspection times $t_{insp,1}$ and $t_{insp,2}$ are 11.78 years and 19.94 years, respectively, and the objective values are $P_{det} = 0.90$, $E(t_{del}) = 7.79$ years, $P_f = 0.08$, $E(t_{mdl}) = 12.33$ years, $E(t_{life}) = 39.54$ years and $E(C_{lcc}) = \$225{,}604$ (see Figure 7.10(b)). The same inspection times and objective values can be obtained from the solutions $I_{R\text{-}T,1}$ and $I_{C\text{-}T,1}$.

7.3.5 Application to Optimum Monitoring Planning

The optimum monitoring planning by the decision making after solving MOLCO is illustrated with the ship hull structure presented in Sub-sections 3.4.4 and 7.2.3. The five objectives $O_{M,1}$, $O_{M,2}$, $O_{M,3}$, $O_{M,4}$ and $O_{M,5}$ are considered simultaneously for given number of monitorings $N_{mon} = 2$ and monitoring duration $t_{md} = 0.3$ year. Through the dominance relation-based

Table 7.6 Optimum inspection times for $N_{insp} = 2$ based on SAW and TOPSIS methods.

Weight determination methods	Weight factors of the objectives $\mathbf{w} = \{w_1, w_2, w_3, w_4, w_5, w_6\}$	MADM methods		Selected Pareto optimal solutions: Optimum inspection times (years)		
SD	{0.0, 0.25, 0.0, 0.17, 0.25, 0.33}	SAW		$I_{S\text{-}S,1}$	$I_{S\text{-}S,2}$	$I_{S\text{-}S,3}$
			$t_{insp,1}$	13.15	13.40	12.58
			$t_{insp,2}$	19.39	19.04	17.23
		TOPSIS		$I_{S\text{-}T,1}$	$I_{S\text{-}T,2}$	$I_{S\text{-}T,3}$
			$t_{insp,1}$	11.78	11.67	11.26
			$t_{insp,2}$	19.94	19.22	19.19
CRITIC	{0.0, 0.23, 0.0, 0.16, 0.30, 0.31}	SAW		$I_{R\text{-}S,1}$	$I_{R\text{-}S,2}$	$I_{R\text{-}S,3}$
			$t_{insp,1}$	13.15	13.40	11.78
			$t_{insp,2}$	19.39	19.04	19.94
		TOPSIS		$I_{R\text{-}T,1}$	$I_{R\text{-}T,2}$	$I_{R\text{-}T,3}$
			$t_{insp,1}$	11.78	11.26	11.67
			$t_{insp,2}$	19.94	19.19	19.22
CCSD	{0.0, 0.17, 0.0, 0.18, 0.29, 0.37}	SAW		$I_{C\text{-}S,1}$	$I_{C\text{-}S,2}$	$I_{C\text{-}S,3}$
			$t_{insp,1}$	13.15	13.40	14.23
			$t_{insp,2}$	19.39	19.04	19.27
		TOPSIS		$I_{C\text{-}T,1}$	$I_{C\text{-}T,2}$	$I_{C\text{-}T,3}$
			$t_{insp,1}$	11.78	13.15	11.67
			$t_{insp,2}$	19.94	19.39	19.22

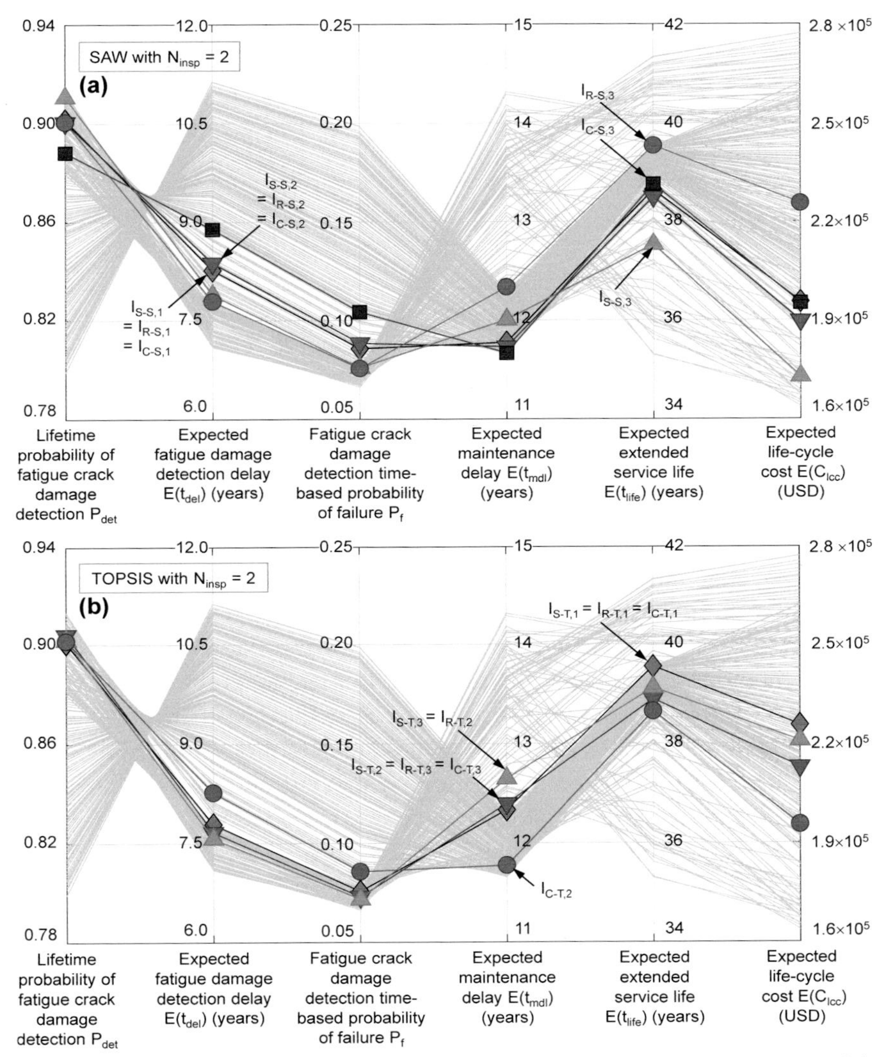

Figure 7.10 Optimum solutions for UL inspection planning with N_{insp} = 2 in the parallel coordinate system: (a) SAW; (b) TOPSIS.

Color version at the end of the book

Table 7.7 Optimum monitoring starting times for $N_{mon} = 2$ and $t_{md} = 0.3$ year based on ELECTRE method.

Weight determination methods	Weight factors of the objectives $\mathbf{w} = \{w_1, w_2, w_3, w_4, w_5\}$	Selected Pareto optimal solutions			
			Optimum monitoring starting times (years)		
			$M_{S\text{-}E,1}$	$M_{S\text{-}E,2}$	$M_{S\text{-}E,3}$
SD	{0.25, 0.25, 0.22, 0.28, 0.0}	$t_{ms,1}$	4.64	4.23	4.17
		$t_{ms,2}$	10.10	9.85	9.39
			$M_{R\text{-}E,1}$	$M_{R\text{-}E,2}$	$M_{R\text{-}E,3}$
CRITIC	{0.20, 0.20, 0.17, 0.42, 0.0}	$t_{ms,1}$	6.18	6.37	6.42
		$t_{ms,2}$	12.68	13.03	13.07
			$M_{C\text{-}E,1}$	$M_{C\text{-}E,2}$	$M_{C\text{-}E,3}$
CCSD	{0.26, 0.27, 0.12, 0.36, 0.0}	$t_{ms,1}$	4.64	4.78	4.23
		$t_{ms,2}$	10.10	11.44	9.85

objective reduction approach, the essential objectives for $N_{mon} = 2$ and $t_{md} = 0.3$ year are identified as $O_{M,1}$, $O_{M,2}$, $O_{M,3}$ and $O_{M,4}$. The weight factors of these essential objectives are determined using the SD, CRITIC and CCSD methods as indicated in Table 7.7. For each weight determination method, the three Pareto optimal solutions associated with the first, second and third largest overall assessment values are determined according to the ELECTRE method. The optimum monitoring starting times (i.e., $t_{ms,1}$ and $t_{ms,2}$) and the corresponding objective values (i.e., $E(t_{del})$, P_f, $E(t_{mdl})$, $E(t_{life})$ and $E(C_{lcc})$) are provided in Table 7.7 and Figure 7.11, respectively.

For example, the solution $M_{S\text{-}E,1}$ has the largest overall assessment value based on the ELECTRE method, when the weight factors $\mathbf{w}$ = {0.25, 0.25, 0.22, 0.28, 0.0} are applied. These weight factors $\mathbf{w}$ are computed using the SD method. The weight factor w_5 for the objective $O_{M,5}$ is equal to zero, since $O_{M,5}$ is redundant. The solution $M_{S\text{-}E,1}$ requires the first monitoring starting time $t_{ms,1} = 4.64$ years and the second monitoring starting time $t_{ms,2} = 10.10$ years, and as a result, $E(t_{del})$, P_f, $E(t_{mdl})$, $E(t_{life})$ and $E(C_{lcc})$ are 2.90 years, 0.05, 7.78 years, 23.43 years and \$371,027, respectively, as shown in Table 7.7 and Figure 7.11(a). If the weight factors are determined by the CCSD method, the solution $M_{C\text{-}E,1}$ with the first largest overall assessment value can be obtained. The monitoring starting times and objective values

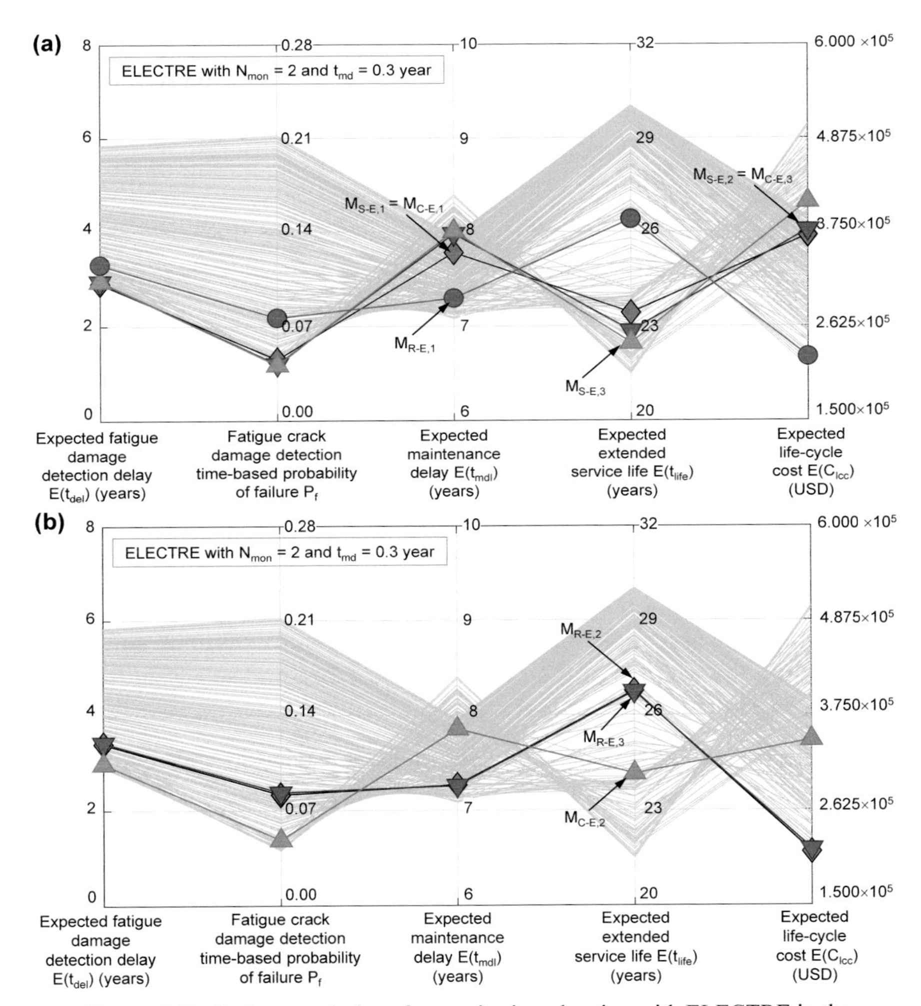

Figure 7.11 Optimum solutions for monitoring planning with ELECTRE in the parallel coordinate system: (a) $M_{S\text{-}E,1} = M_{C\text{-}E,1}$, $M_{S\text{-}E,2} = M_{C\text{-}E,3}$, $M_{R\text{-}E,1}$, $M_{S\text{-}E,3}$; (b) $M_{R\text{-}E,2}$, $M_{R\text{-}E,3}$, $M_{C\text{-}E,2}$.

Color version at the end of the book

for $M_{C\text{-}E,1}$ are the same as those for $M_{S\text{-}E,1}$. The solution $M_{C\text{-}E,2}$ has the second largest overall assessment value. The associated monitoring starting times $t_{ms,1}$ and $t_{ms,2}$ are 4.78 years and 11.44 years, respectively. By applying these monitoring starting times, $E(t_{del})$ = 2.98 years, P_f = 0.05, $E(t_{mdl})$ = 7.86 years, $E(t_{life})$ = 24.15 years and $E(C_{lcc})$ = \$345,576 can be expected as shown in Table 7.7 and Figure 7.11(b).

7.4 Conclusions

In this chapter, the multi-objective decision making framework for optimum inspection and monitoring planning is investigated. The presented framework consists of two decision alternatives to select the best single optimum inspection and monitoring plan: decision making *before* and *after* solving MOLCO. The decision making before solving MOLCO is performed to determine the weight factors of the objectives, and to convert the multi-objectives into a single objective. The single optimum inspection and monitoring plan can be established by solving the converted single objective optimization without obtaining the entire Pareto optimal solution set. The decision making after MOLCO includes identifying the essential objectives with the Pareto optimal solutions, determining the weights of the essential objectives, and selecting the best solution from the Pareto optimal set.

The selection of the best optimal solution for inspection and monitoring planning is based on various weight determination methods and MADM methods. In this chapter, the subjective weight determination methods (i.e., ranking, rating and paired comparison methods) for decision making before solving MOLCO and the objective weight determination methods (i.e., SD, CRITIC and CCSD methods) for decision making after solving MOLCO are used. To establish the best optimal inspection and monitoring plan, MADM methods such as the SAW, TOPSIS, and ELECTRE methods are applied.

The decision making before solving MOLCO can be more cost-efficient than the decision making after MOLCO. This is because the decision making after MOLCO requires the Pareto optimal set. The multi-objective optimization process for finding the Pareto optimal set generally requires more computational cost. However, the decision making after MOLCO can be more flexible in determining the best Pareto optimal solution than that before MOLCO.

Chapter **8**

Conclusions

CONTENTS

ABSTRACT

Finally, the summary of this book, representative conclusions derived from each chapter, and future directions are presented in Chapter 8.

8.1 Summary

Life-cycle optimum service life management integrates damage propagation, damage detection, inspection, monitoring, maintenance, probability of failure, service life, and total life-cycle cost under uncertainty. Such an integration is addressed in this book with an emphasis on deteriorating civil and marine structures under fatigue. Chapter 1 describes the general concepts of life-cycle analysis and optimization under uncertainty. In Chapter 2, the role of inspection and monitoring in life-cycle analysis is presented. Chapter 3 deals with the probability of fatigue crack damage detection. The concepts and approaches for damage detection-based optimum inspection and monitoring planning are addressed in Chapter 4. Chapter 5 deals with the optimum inspection and monitoring planning, considering the effects of inspection, monitoring and maintenance on service life extension and life-cycle cost. Chapter 6 presents the multi-objective life-cycle optimization using the objectives formulated in Chapters 4 and 5. For practical applications, decision making with the Pareto solutions from the multi-objective life-cycle optimization are presented in Chapter 7.

8.2 Conclusions

The following conclusions can be drawn:

- Life-cycle structural performance and life-cycle cost analysis and prediction are essential for optimum service life management of deteriorating structures. In order to perform this optimization, appropriate understanding of (a) structural performance deterioration subjected to corrosion and fatigue, (b) effects of maintenance actions on structural performance, cost and service life, and (c) probabilistic structural performance indicators is required.
- Accuracy of the structural performance assessment and prediction can be improved through the efficient use of inspection and monitoring data. This improvement results in appropriate and timely maintenance, and prevention of the unexpected failure of a structure. Statistical and probabilistic concepts regarding (a) availability of monitoring data,

(b) monetary loss caused by use of unavailable inspection and monitoring data, (c) updating processes based on information from inspection and monitoring, and (d) importance indicators of the individual components in a structural system, help managers to use efficient inspection and monitoring planning for structural components and systems.

- Time-dependent fatigue crack propagation affects the probability of fatigue damage detection when inspections and/or monitorings are performed. The relation between the fatigue crack propagation and the probability of fatigue damage detection is used to formulate and estimate the probability of fatigue crack damage detection, expected damage detection time and delay, and damage detection time-based probability of failure. It can be concluded that an increase in the lifetime probability of fatigue crack damage detection reduces both the expected damage detection delay and the damage detection time-based probability of failure.
- Single-objective optimum inspection and monitoring planning for service life management of fatigue-sensitive structures can be based on maximizing the lifetime probability of fatigue crack damage detection, or minimizing the expected fatigue crack damage detection delay, or minimizing the fatigue crack damage detection time-based probability of failure. By solving the single-objective optimizations associated with these separate objectives, the optimum inspection and monitoring application times can be obtained. The results presented reveal that an increase in the number of inspections and/or an improvement of the inspection quality can lead to (a) earlier inspection time, (b) increase in the probability of fatigue damage detection, and (c) reduction of both the expected fatigue crack damage detection delay and damage detection time-based probability of failure. Furthermore, the monitoring planning associated with a large number of monitorings and long monitoring duration reduces both the expected fatigue crack damage detection delay and the fatigue crack damage detection time-based probability of failure.
- The effect of inspection and maintenance on the service life extension and life-cycle cost is implemented in the formulation of the objectives: minimizing the expected maintenance delay, or maximizing the expected

extended service life, or minimizing the expected life-cycle cost. The results of the single-objective optimizations associated with these separate objectives show that (a) the number and type of inspections, (b) the number and duration of monitorings, (c) critical fatigue crack size for maintenance actions, and (d) monetary loss due to failure will affect the optimum life-cycle management of deteriorating civil and marine structures.

- A single-objective optimization can provide inspection and/or monitoring plans by considering the objectives separately. When multiple objectives for optimum life-cycle management are available, all the objectives can be integrated and the multiple trade-off solutions can be obtained through multi-objective optimization. However, as the number of objectives increases, the dimensionality to express the Pareto optimal set increases, and, therefore, there will be difficulties associated with both the illustration of the Pareto solutions in the 3D Cartesian coordinate system and the decision making process to select the best-balanced solution from the Pareto optimal set.
- The multi-objective decision making framework presented is able to select the best-balanced optimum inspection and monitoring plan from the Pareto optimal set. The presented decision making framework has two alternatives: decision making before and after solving multi-objective life-cycle optimization (MOLCO). The decision making before solving MOLCO is performed to convert the multi-objectives into a single objective by determining the weight factors of the objectives and to establish a single optimum inspection and monitoring plan without finding the entire Pareto optimal solution set. The decision making after MOLCO needs the Pareto optimal solutions to select the best solution from the Pareto optimal set. In general, the multi-objective optimization process requires large computational costs to find the Pareto optimal set. For this reason, the decision making before solving MOLCO can be more efficient than the decision making after MOLCO in terms of computational cost. However, the decision making after MOLCO can be more flexible in determining the best-balanced Pareto optimal solution.

8.3 Future Directions

This book provides fundamentals and recent achievements in life-cycle management of fatigue-sensitive structures. However, there are still further developments to be addressed, as follows:

- The results of life-cycle management of civil and marine structures highly depend on the accuracy of the performance prediction process. The practical application of life-cycle management requires accurate information on the structural performance and the associated probabilistic parameters. Although modeling approaches and techniques for the structural performance prediction and integration of structural health monitoring (SHM) information for the efficient life-cycle management have been proposed, accurate methodologies for continuous updating based on a large amount of data and long-term SHM are needed.
- The failure cost representing the consequences of structural failure significantly affects the life-cycle management of deteriorating structures. A deteriorating structure with a high failure cost requires a large number of inspections, monitorings and maintenance actions (Frangopol et al. 1997; Frangopol and Kim 2011; Zhu and Frangopol 2013a). For this reason, a rational estimation of the failure cost considering economic, social and environmental losses is needed.
- In order to formulate the objective functions for the optimum inspection planning, the relationship between the probability of detection and fatigue crack size needs to be quantified as a function of the probability of fatigue damage detection. The probability of damage detection for an inspection method has to be formulated based on extensive field data under uncertainty. Further efforts are still required to improve the accuracy of the probability of damage detection of inspection methods used for fatigue-sensitive structures.
- The formulation of the objectives for monitoring planning are based on the assumption that fatigue crack damage is detected perfectly during the monitoring period. The uncertainty associated with fatigue damage detection during the monitoring period can be caused by inaccurate SHM data, and/or inappropriate interpretation of SHM

data. Further studies for the life-cycle management of deteriorating structures need to consider this uncertainty.

- The structural performance improvement and service life extension through maintenance actions are related to the type of applied maintenance, degree and location of damage, and contribution of a maintained component to the reliability of the structural system. Further research is required to model the effects of maintenance on structural performance and service life extension.
- Life-cycle analysis, prediction and optimum management of deteriorating civil and marine structures are, in general, not applied extensively in practice, even though advanced probabilistic techniques and approaches have been developed. This situation is due to the lack of validation of the proposed techniques and approaches. Therefore, future efforts should place more emphasis on real-world application to civil and marine structures by developing specifications and guidelines for life-cycle probabilistic analysis, prediction and optimum management of fatigue-sensitive structures.

References

Akpan, U.O., Koko, T.S., Ayyub, B. and Dunbar, T.E. 2002. Risk assessment of aging ship hull structures in the presence of corrosion and fatigue. Marine Structures, Elsevier, 15(3): 211–231.

Akiyama, M., Frangopol, D.M. and Matsuzaki, H. 2011. Life-cycle reliability of RC bridge piers under seismic and airborne chloride hazards. Earthquake Engineering & Structural Dynamics, John Wiley & Sons, Ltd., 40(15): 1671–1687.

Akiyama, M., Frangopol, D.M., Arai, M. and Koshimura, S. 2013. Reliability of bridges under tsunami hazard: Emphasis on the 2011 Great East Japan earthquake. Earthquake Spectra, EERI, 29(S1): S295–S314.

Ang, A.H.-S. and Tang, W.H. 1984. Probability Concepts in Engineering Planning and Design Volume II. John Wiley & Sons.

Ang, A.H.-S. and Tang, W.H. 2007. Probability Concepts in Engineering: Emphasis on Applications to Civil and Environmental Engineering. 2nd Edition. New York, Wiley.

Antonaci, P., Bocca, P. and Masera, D. 2012. Fatigue crack propagation monitoring by Acoustic Emission signal analysis. Engineering Fracture Mechanics, Elsevier, 81: 26–32.

Arora, J.S. 2016. Introduction to Optimum Design. 4th Edn. Elsevier, UK.

Arora, P., Popov, B.N., Haran, B., Ramasubramanian, M., Popova, S. and White, R.E. 1997. Corrosion initiation time of steel reinforcement in a chloride environment—A one dimensional solution. Corrosion Science, Elsevier, 39(4): 739–759.

ASCE. 2017. Report Card for America's Infrastructure. American Society of Civil Engineers, Reston, VA.

Baker, J.W., Schubert, M. and Faber, M.H. 2008. On the assessment of robustness. Structural Safety. Elsevier, 30: 253–267.

Barone, G. and Frangopol, D.M. 2013. Hazard-based optimum lifetime inspection and repair planning for deteriorating structures. Journal of Structural Engineering, ASCE, 139(12): 04013017,1-12.

Barone, G. and Frangopol, D.M. 2014. Reliability, risk and lifetime distributions as performance indicators for life-cycle maintenance of deteriorating structures. Reliability Engineering & System Safety, Elsevier, 123(3): 21–37.

Barone, G., Frangopol, D.M. and Soliman, M. 2014. Optimization of life-cycle maintenance of deteriorating structures considering expected annual system failure rate and expected cumulative cost. Journal of Structural Engineering, ASCE, 140(2): 04013043, 1-13.

Berens, A.P. and Hovey, P.W. 1981. Evaluation of NDE reliability characterization. Air Force Wright-Aeronautical Laboratory, Wright-Patterson Air Force Base, Dayton, Ohio.

Berens, A.P. 1989. NDE reliability analysis. Metal handbook, 9th Edition, Vol. 17, ASM International, Material Park, Ohio, 689–701.

Biondini, F. and Frangopol, D.M. 2016. Life-cycle performance of deteriorating structural systems under uncertainty: Review. Journal of Structural Engineering, ASCE, 142(9): F4016001, 1–17.

Birnbaum, Z.W. 1969. On the importance of different components in a multicomponent system. *In*: Krishnaiah, P.R. (ed.). Multivariate Analysis-II, Academic Press, New York.

Bocchini, P. and Frangopol, D.M. 2011. A probabilistic computational framework for bridge network optimal maintenance scheduling. Reliability Engineering & System Safety, Elsevier, 96(2): 332–349.

Brockhoff, D. and Zitzler, E. 2006. Dimensionality reduction in multiobjective optimization with (partial) dominance structure preservation: Generalized minimum objective subset problems. TIK Report 247, ETH Zurich, Switzerland.

Brockhoff, D. and Zitzler, E. 2007. Improving hypervolume-based multiobjective evolutionary algorithms by using objective reduction methods. IEEE Congress on Evolutionary Computation, IEEE Press, Singapore, pp. 2086–2093.

Brockhoff, D. and Zitzler, E. 2009. Objective reduction in evolutionary multiobjective optimization: Theory and applications. Evolutionary Computation, MIT Press, 17(2): 135–166.

Brownjohn, J.M.W. 2007. Structural health monitoring of civil infrastructure. Philosophical Transactions of the Royal Society A, Royal Society Publishing, 365(1851): 589–622.

Bucher, C. and Frangopol, D.M. 2006. Optimization of lifetime maintenance strategies for deteriorating structures considering probabilities of violating safety, condition, and cost thresholds. Probabilistic Engineering Mechanics, Elsevier, 21(1): 1–8.

Bucher, C. 2009. Computational Analysis of Randomness in Structural Mechanics: Vol. 3. Structures and Infrastructures Series. Frangopol, D.M. (ed.). Leiden, The Netherlands: CRC Press/Balkema/Taylor & Francis.

Cambridge Systematics, Inc. 2009. Pontis Release 4.5 User Manual, AASHTO, Washington, DC.

Catbas, F.N., Susoy, M. and Frangopol, D.M. 2008. Structural health monitoring and reliability estimation: Long span truss bridge application with environmental monitoring data. Engineering Structures, Elsevier, 30(9): 2347–2359.

Catbas, N., Gokce, H.B. and Frangopol, D.M. 2013. Incorporating uncertainty through a family of models calibrated with structural health monitoring data for reliability prediction. Journal of Engineering Mechanics, ASCE, 139(6): 712–723.

Cho, H. and Lissenden, C.J. 2012. Structural health monitoring of fatigue crack growth in plate structures with ultrasonic guided waves. Structural Health Monitoring, SAGE, 11(4): 393–404.

Chung, H.-Y., Manuel, L. and Frank, K.H. 2006. Optimal inspection scheduling of steel bridges using nondestructive testing techniques. Journal of Bridge Engineering, ASCE, 11(3): 305–319.

Ciang, C.C., Lee, J.-R. and Bang, H.-J. 2008. Structural health monitoring for a wind turbine system: a review of damage detection methods. Measurement Science and Technology, IOP Science, 19(12): 122001.

Colorado Department of Transportation (CDOT). 1998. Pontis Bridge Inspection Coding Guide, Denver, Colorado.

Connor, R.J. and Fisher, J.W. 2001. Report on field measurements and assessment of the I-64 Kanawha River Bridge at Dunbar, West Virginia. Report No. 01–14, Lehigh University's Center for Advanced Technology for Large Structural Systems (ATLSS), Bethlehem, PA.

Connor, R.J. and Lloyd, J.B. 2017. Maintenance Actions to Address Fatigue Cracking in Steel Bridge Structures: Proposed Guidelines and Commentary. West Lafayette, IN: Purdue University.

Cornell, C.A. 1967. Bounds on the reliability of structural systems. Journal of Structural Division, ASCE, 93(ST1): 171–200.

Crawshaw, J. and Chambers, J. 1984. A Concise Course in A-Level Statistics. Stanley Thornes (Publishers) Ltd.

Deb, K. 2001. Multi-Objective Optimization Using Evolutionary Algorithms. John Wiley & Sons, New York.

Deb, K. and Saxena, D. 2006. Searching for Pareto-optimal solutions through dimensionality reduction for certain large-dimensional multi-objective optimization problems. Proceedings of the IEEE Congress on Evolutionary Computation (CEC2006), July 16–21, Vancouver, Canada.

Decò, A. and Frangopol, D.M. 2011. Risk assessment of highway bridges under multiple hazards. Journal of Risk Research, Taylor & Francis, 14(9): 1057–1089.

Decò, A., Frangopol, D.M. and Okasha, N.M. 2011. Time-variant redundancy of ship structures. Journal of Ship Research, SNAME, 55(3): 208–219.

Decò, A., Frangopol, D.M. and Zhu, B. 2012. Reliability and redundancy assessment of ships under different operational conditions. Engineering Structures, Elsevier, 42(9): 457–471.

Decò, A. and Frangopol, D.M. 2013. Life-cycle risk assessment of spatially distributed aging bridges under seismic and traffic hazards. Earthquake Spectra, EERI, 29(1): 127–153.

Decò, A. and Frangopol, D.M. 2015. Real-time risk of ship structures integrating structural health monitoring data: Application to multi-objective optimal ship routing. Ocean Engineering, Elsevier, 96: 312–329.

Demsetz, L., Cario, R. and Schulte-Strathaus, R. 1996. Inspection of Marine Structures. Ship Structures Committee Report No. SSC-389. Washington, DC: Ship Structures Committee.

Deng, H., Yeh, C.-H. and Willis, R.J. 2000. Inter-company comparison using modified TOPSIS with objective weights. Computers & Operations Research, Pergamon, 27(10): 963–973.

Dexter, R.J., FitzPatrick, R.J. and St. Peter, D.L. 2003. Fatigue strength and adequacy of weld repairs. Washington, DC, Ship Structure Committee, Report No. SSC-425.

Diakoulaki, D., Mavrotas, G. and Papayannakis, L. 1995. Determining objective weights in multiple criteria problems: The critic method. Computers & Operations Research, Pergamon, 22(7): 763–770.

Ditlevsen, O. 1979. Narrow reliability bounds for structural systems. Journal of Structural Mechanics, ASCE, 7(4): 453–472.

Dong, Y. and Frangopol, D.M. 2015. Risk-informed life-cycle optimum inspection and maintenance of ship structures considering corrosion and fatigue. Ocean Engineering, Elsevier, 101: 161–171.

Dong, Y. and Frangopol, D.M. 2016. Incorporation of risk and updating in inspection of fatigue-sensitive details of ship structures. International Journal of Fatigue, Elsevier, Part 3, 82: 676–688.

Dong, Y. and Frangopol, D.M. 2017. Adaptation optimization of residential buildings under hurricane threat considering climate change in a lifecycle context. Journal of Performance of Constructed Facilities, ASCE, 31(6): 4017099.

Ellingwood, B.R. and Mori, Y. 1997. Reliability-based service life assessment of concrete structures in nuclear power plants: optimum inspection and repair. Nuclear Engineering and Design, Elsevier, 175(3): 247–258.

Ellingwood, B.R. 2006. Mitigating risk from abnormal loads and progressive collapse. Journal of Performance of Constructed Facilities, ASCE, 20(4): 315–323.

Enright, M.P. and Frangopol, D.M. 1999a. Maintenance planning for deteriorating concrete bridges. Journal of Structural Engineering, ASCE, 125(12): 1407–1414.

Enright, M.P. and Frangopol, D.M. 1999b. Condition prediction of deteriorating concrete bridges using Bayesian updating. Journal of Structural Engineering, ASCE, 125(10): 1118–1124.

Ericson, C. 2015. Hazard Analysis Techniques for System Safety. 2nd Edition. New York, Wiley.

Estes, A.C. and Frangopol, D.M. 1999. Repair optimization of highway bridges using system reliability approach. Journal of Structural Engineering, ASCE, 125(7): 766–775.

Estes, A.C. and Frangopol, D.M. 2001. Minimum expected cost-oriented optimal maintenance planning for deteriorating structures: application to concrete bridge decks. Reliability Engineering & System Safety, Elsevier, 73(3): 281–291.

Estes, A.C. and Frangopol, D.M. 2003. Updating bridge reliability based on bridge management systems visual inspection results. Journal of Bridge Engineering, ASCE, 8(6): 374–382.

Farhey, D.N. 2005. Bridge instrumentation and monitoring for structural diagnostics. Structural Health Monitoring, SAGE, 4(4): 301–318.

Fatemi, A. and Yang, L. 1998. Cumulative fatigue damage and life prediction theories: a survey of the state of the art for homogeneous materials. International Journal of Fatigue, 20(1): 9–34.

FHWA. 1995. Recording and Coding Guide for Structure Inventory and Appraisal of the Nation's Bridge, Report No. FHWA-PD 96-001, U.S. Department of Transportation, Washington, DC.

FHWA. 2013. Manual for Repair and Retrofit of Fatigue Cracks in Steel Bridges. FHWA Publication No. FHWA-IF-13-020, Arlington, VA, USA.

Fisher, J.W. 1984. Fatigue and Fracture in Steel Bridges. Wiley, New York.

Fisher, J.W., Kulak, G.L. and Smith, I.F. 1998. A Fatigue Primer for Structural Engineers. National Steel Bridge Alliance, Chicago, IL, USA.

Flintsch, G. and Chen, C. 2004. Soft computing applications in infrastructure management. Journal of Infrastructure Systems, 10(4): 157–166.

Fonseca, C.M. and Fleming, P.J. 1998. Multiobjective optimization and multiple constraint handling with evolutionary algorithms. I. A Unified Formulation, in Systems, Man and Cybernetics, Part A: Systems and Humans, 28(1): 26–37.

Forsyth, D.S. and Fahr, A. 1998. An evaluation of probability of detection statistics. RTO-AVT Workshop on Airframe inspection reliability under field/depot conditions, Brussels, Belgium, pp. 10.1–10.5.

Frangopol, D.M. and Curley, J.P. 1987. Effects of damage and redundancy on structural reliability. Journal of Structural Engineering, ASCE, 113(7): 1533–1549.

Frangopol, D.M. and Nakib, R. 1991. Redundancy in highway bridges. Engineering Journal, American Institute of Steel Construction (AISC), Chicago, IL, 28(1): 45–50.

Frangopol, D.M. and Estes, A.C. 1997. Lifetime bridge maintenance strategies based on system reliability. Structural Engineering International, IABSE, 7(3): 193–198.

Frangopol, D.M., Lin, K.Y. and Estes, A.C. 1997. Life-cycle cost design of deteriorating structures. Journal of Structural Engineering, ASCE, 123(10): 1390–1401.

Frangopol, D.M., Gharaibeh, E.S., Kong, J.S. and Miyake, M. 2000. Optimal network-level bridge maintenance planning based on minimum expected cost. Journal of the Transportation Research Board, Transportation Research Record, 1696(2), National Academy Press: 26–33.

Frangopol, D.M. and Maute, K. 2003. Life-cycle reliability-based optimization of civil and aerospace structures. Computers & Structures, Elsevier, 81(7): 397–410.

Frangopol, D.M., Kallen, M-J. and van Noortwijk, J. 2004. Probabilistic models for life-cycle performance of deteriorating structures: review and future directions. Progress in Structural Engineering and Materials, John Wiley & Sons, 6(4): 197–212.

Frangopol, D.M. and Liu, M. 2007. Maintenance and management of civil infrastructure based on condition, safety, optimization, and life-cycle cost. Structure and Infrastructure Engineering, Taylor & Francis, 3(1): 29–41.

Frangopol, D.M., Strauss, A. and Kim, S. 2008a. Bridge reliability assessment based on monitoring. Journal of Bridge Engineering, ASCE, 13(3): 258–270.

Frangopol, D.M., Strauss, A. and Kim, S. 2008b. Use of monitoring extreme data for the performance prediction of structures: General approach. Engineering Structures, Elsevier, 30 (12): 3644–3653.

Frangopol, D.M. 2011. Life-cycle performance, management, and optimization of structural systems under uncertainty: accomplishments and challenges. Structure and Infrastructure Engineering, Taylor & Francis, 7(6): 389–413.

Frangopol, D.M. and Messervey, T.B. 2011. Effect of monitoring on reliability of structures. Chapter 18. pp. 515–560. *In*: Bakht, B., Mufti, A.A. and Wegner, L.D. (eds.). Monitoring Technologies for Bridge Management, Multi-Science Publishing Co. Ltd. U.K.

Frangopol, D.M. and Kim, S. 2011. Service life, reliability and maintenance of civil structures. Chapter 5. pp. 145–178. *In*: Lee, L.S. and Karbari, V. (eds.). Service Life Estimation and Extension of Civil Engineering Structures. Woodhead Publishing Ltd., Cambridge, U.K.

Frangopol, D.M., Bocchini, P., Decò, A., Kim, S., Kwon, K., Okasha, N.M. and Saydam, D. 2012. Integrated life-cycle framework for maintenance, monitoring, and reliability of naval ship structures. Naval Engineers Journal, Wiley, 124(1): 89–99.

Frangopol, D.M., Saydam, D. and Kim, S. 2012. Maintenance, management, life-cycle design and performance of structures and infrastructures: A brief review. Structure and Infrastructure Engineering, Taylor & Francis, 8(1): 1–25.

Frangopol, D.M. and Saydam, D. 2014. Structural performance indicators for bridges. Chapter 9. pp. 185–206. *In*: Chen, W.-F. and Duan, L. (eds.). Bridge Engineering Handbook. Second Edition, Vol. 1 Fundamentals, CRC Press/Taylor & Francis Group, Boca Raton, London, New York.

Frangopol, D.M. and Kim, S. 2014a. Bridge health monitoring. Chapter 10. pp. 247–268. *In*: Chen, W.-F. and Duan, L. (eds.). Bridge Engineering Handbook. Second Edition, Vol. 5 Construction and Maintenance, CRC Press/Taylor & Francis Group, Boca Raton, London, New York.

Frangopol, D.M. and Kim, S. 2014b. Life-cycle analysis and optimization. Chapter 18. pp. 537–566. *In*: Chen, W.-F. and Duan, L. (eds.). Bridge Engineering Handbook. Second Edition, Vol. 5 Construction and Maintenance, CRC Press/Taylor & Francis Group, Boca Raton, London, New York.

Frangopol, D.M. and Kim, S. 2014c. Prognosis and life-cycle assessment based on SHM information. Chapter 5. pp. 145–171. *In*: Wang, M.L., Lynch, J. and Sohn, H. (eds.). Part II. Data Interrogation and Decision Making in Sensor Technologies for Civil Infrastructures: Performance Assessment and Health Monitoring, Woodhead Publishing Ltd., Cambridge.

Frangopol, D.M. and Soliman, M. 2016. Life-cycle of structural systems: recent achievements and future directions. Structure and Infrastructure Engineering, Taylor & Francis, 12(1): 1–20.

Frangopol, D.M., Dong, Y. and Sabatino, S. 2017. Bridge life-cycle performance and cost: Analysis, prediction, optimization, and decision-making. Structure and Infrastructure Engineering, Taylor & Francis, 13(10): 1239–1257.

Frangopol, D.M. 2018. Structures and Infrastructure Systems: Life-Cycle Performance, Management, and Optimization. Routledge, 406 pages.

Fu, G. and Frangopol, D.M. 1990. Balancing weight, system reliability and redundancy in a multiobjective optimization framework. Structural Safety, Elsevier, 7(2-4): 165–175.

Fussell, J. 1975. How to hand calculate system reliability and safety characteristics. IEEE Transactions on Reliability, IEEE, R-24(3): 169–174.

Garbatov, Y. and Guedes Soares, C. 2001. Cost and reliability based strategies for fatigue maintenance planning of floating structures. Reliability Engineering & System Safety, Elsevier, 73(3): 293–301.

Garbatov, Y. and Guedes Soares, C. 2014. Risk-based Maintenance of Aging Ship Structures. Maintenance and Safety of Aging Infrastructure: Structures and Infrastructures Book Series. Frangopol, D.M. and Tsompanakis, Y. (eds.). CRC Press/Balkema—Taylor & Francis Group, London, Chapter 11, 10: 307–342.

Ghosn, M., Moses, F. and Frangopol, D.M. 2010. Redundancy and robustness of highway bridge superstructures and substructures. Structure and Infrastructure Engineering, Taylor & Francis, 6(1-2): 257–278.

Ghosn, M., Dueñas-Osorio, L., Frangopol, D.M., McAllister, T.P., Bocchini, P., Manuel L., Ellingwood, B.R., Arangio, S., Bontempi, F., Shah, M., Akiyama, M., Biondini, F., Hernandez, S. and Tsiatas, G. 2016a. Performance indicators for structural systems and infrastructure networks. Journal of Structural Engineering, ASCE, 142(9): F4016003, 1–18.

Ghosn, M., Frangopol, D.M., McAllister, T.P., Shah, M., Diniz, S., Ellingwood, B.R., Manuel, L., Biondini, F., Catbas, N., Strauss, A. and Zhao, Z.L. 2016b. Reliability-based structural performance indicators for structural members. Journal of Structural Engineering, ASCE, 142(9): F4016002, 1–13.

Glen, L.F., Dinovitzer, A., Malik, L., Basu, R. and Yee, R. 2000. Guide to damage tolerance analysis of marine structures. Rep. No. SSC-409, Ship Structure Committee, Washington, DC.

Gonzalez, J.A., Andrade, C., Alonso, C. and Feliu, S. 1995. Comparison of rates of general corrosion and maximum pitting penetration on concrete embedded steel reinforcement. Cement and Concrete Research, Elsevier, 25(2): 257–264.

Guedes Soares, C. and Garbatov, Y. 1996a. Fatigue reliability of the ship hull girder accounting for inspection and repair. Reliability Engineering & System Safety, Elsevier, 51(3): 341–351.

Guedes Soares, C. and Garbatov, Y. 1996b. Fatigue reliability of the ship hull girder. Marine Structures, Elsevier, 9(3-4): 495–516.

Gumbel, E.J. 1958. Statistics of Extremes. Columbia Univ. Press, New York, NY.

Hellier, C.J. 2012. Handbook of Nondestructive Evaluation. 2nd edn., McGraw-Hill, New York, NY.

Hendawi, S. and Frangopol, D.M. 1994. System reliability and redundancy in structural design and evaluation. Structural Safety, Elsevier, 16(1-2): 47–71.

Hoyland, A. and Rausand, M. 1994. System Reliability Theory: Models and Statistical Methods, Wiley, New York.

Hwang, C.-L., Lai, Y.-J. and Liu, T.-Y. 1993. A new approach for multiple objective decision making. Computers & Operations Research, Pergamon, 20(8): 889–899.

Irwin, G.R. 1958. The crack-extension-force for a crack at a free surface boundary. Washington, DC, Naval Research Laboratory, Report No. 5120.

Kabir, G., Sadiq, R. and Tesfamariam, S. 2014. A review of multi-criteria decision-making methods for infrastructure management. Structure and Infrastructure Engineering, 10(9): 1176–1210.

Kim, S. and Frangopol, D.M. 2010. Optimal planning of structural performance monitoring based on reliability importance assessment. Probabilistic Engineering Mechanics, Elsevier, 25(1): 86–98.

Kim, S., Frangopol, D.M. and Zhu, B. 2011. Probabilistic optimum inspection/repair planning to extend lifetime of deteriorating structures. Journal of Performance of Constructed Facilities, ASCE, 25(6): 534–544.

Kim, S. and Frangopol, D.M. 2011a. Cost-based optimum scheduling of inspection and monitoring for fatigue-sensitive structures under uncertainty. Journal of Structural Engineering, ASCE, 137(11): 1319–1331.

Kim, S. and Frangopol, D.M. 2011b. Cost-effective lifetime structural health monitoring based on availability. Journal of Structural Engineering, ASCE, 137(1): 22–33.

Kim, S. and Frangopol, D.M. 2011c. Inspection and monitoring planning for RC structures based on minimization of expected damage detection delay. Probabilistic Engineering Mechanics, Elsevier, 26(2): 308–320.

Kim, S. and Frangopol, D.M. 2011d. Optimum inspection planning for minimizing fatigue damage detection delay of ship hull structures. International Journal of Fatigue, Elsevier, 33(3): 448–459.

Kim, S. and Frangopol, D.M. 2012. Probabilistic bicriterion optimum inspection/monitoring planning: applications to naval ships and bridges under fatigue. Structure and Infrastructure Engineering, Taylor & Francis, 8(10): 912–927.

Kim, S., Frangopol, D.M. and Soliman, M. 2013. Generalized probabilistic framework for optimum inspection and maintenance planning. Journal of Structural Engineering, ASCE, 139(3): 435–447.

Kim, S. and Frangopol, D.M. 2017. Efficient multi-objective optimisation of probabilistic service life management. Structure and Infrastructure Engineering, Taylor & Francis, 13(1): 147–159.

Kim, S. and Frangopol, D.M. 2018a. Multi-objective probabilistic optimum monitoring planning considering fatigue damage detection, maintenance, reliability, service life and cost. Structural and Multidisciplinary Optimization, Springer, 57(1): 39–54.

Kim, S. and Frangopol, D.M. 2018b. Decision making for probabilistic fatigue inspection planning based on multi-objective optimization. International Journal of Fatigue, Elsevier, 111: 356–368.

Kong, J.S. and Frangopol, D.M. 2003a. Evaluation of expected life-cycle maintenance cost of deteriorating structures. Journal of Structural Engineering, ASCE, 129(5): 682–691.

Kong, J.S. and Frangopol, D.M. 2003b. Life-cycle reliability-based maintenance cost optimization of deteriorating structures with emphasis on bridges. Journal of Structural Engineering, ASCE, 129(6): 818–828.

Kong, J. and Frangopol, D.M. 2004. Cost-reliability interaction in life-cycle cost optimization of deteriorating structures. Journal of Structural Engineering, ASCE, 130(11): 1704–1712.

Kong, J.S. and Frangopol, D.M. 2005. Probabilistic optimization of aging structures considering maintenance and failure costs. Journal of Structural Engineering, ASCE, 131(4): 600–616.

Kwon, K. and Frangopol, D.M. 2010. Bridge fatigue reliability assessment using probability density functions of equivalent stress range based on field monitoring data. International Journal of Fatigue, Elsevier, 32(8): 1221–1232.

Kwon, K. and Frangopol, D.M. 2011. Bridge fatigue assessment and management using reliability-based crack growth and probability of detection models. Probabilistic Engineering Mechanics, Elsevier, 26(3): 471–480.

Kwon, K. and Frangopol, D.M. 2012. System reliability of ship hull structures under corrosion and fatigue. Journal of Ship Research, SNAME, 56(4): 234–251.

Kwon, K., Frangopol, D.M. and Kim, S. 2013. Fatigue performance assessment and service life prediction of high-speed ship structures based on probabilistic lifetime sea loads. Structure and Infrastructure Engineering, Taylor & Francis, 9(2): 102–115.

Leemis, L.M. 2009. Reliability: Probabilistic Models and Statistical Methods. 2nd Edition. Ascended Ideas, U.S.

Lind, N.C. 1995. A measure of vulnerability and damage tolerance. Reliability Engineering & System Safety, Elsevier, 43(1): 1–6.

Liu, M. and Frangopol, D.M. 2004. Optimal bridge maintenance planning based on probabilistic performance prediction. Engineering Structures, Elsevier, 26(7): 991–1002.

Liu, M. and Frangopol, D.M. 2005a. Time-dependent bridge network reliability: novel approach. Journal of Structural Engineering, ASCE, 131(2): 329–337.

Liu, M. and Frangopol, D.M. 2005b. Multiobjective maintenance planning optimization for deteriorating bridges considering condition, safety and life-cycle cost. Journal of Structural Engineering, ASCE, 131(5): 833–842.

Liu, M. and Frangopol, D.M. 2006. Optimizing bridge network maintenance management under uncertainty with conflicting criteria: Life-cycle maintenance, failure, and user costs. Journal of Structural Engineering, ASCE, 132(11): 1835–1845.

Liu, M., Frangopol, D.M. and Kim, S. 2009a. Bridge safety evaluation based on monitored live load effects. Journal of Bridge Engineering, ASCE, 14(4): 257–269.

Liu, M., Frangopol, D.M. and Kim, S. 2009b. Bridge system performance assessment from structural health monitoring: a case study. Journal of Structural Engineering, ASCE, 135(6): 733–742.

Liu, M., Frangopol, D.M. and Kwon, K. 2010a. Optimization of retrofitting distortion-induced fatigue cracking of steel bridges using monitored data under uncertainty. Engineering Structures, Elsevier, 32(11): 3467–3477.

Liu, M., Frangopol, D.M. and Kwon, K. 2010b. Fatigue reliability assessment of retrofitted steel bridges integrating monitored data. Structural Safety, Elsevier, 32(1): 77–89.

Liu, Y. and Frangopol, D.M. 2019a. Risk-informed structural repair decision making for service life extension of aging naval ships. Marine Structures, Elsevier, 64: 305–321.

Liu, Y. and Frangopol, D.M. 2019b. Utility and information analysis for optimum inspection of fatigue-sensitive structures. Journal of Structural Engineering, ASCE, 145(2): 04018251, 1–12.

Lukić, M. and Cremona, C. 2001. Probabilistic optimization of welded joints maintenance versus fatigue and fracture. Reliability Engineering & System Safety, Elsevier, 72(3): 253–264.

Madsen, H.O., Krenk, S. and Lind, N.C. 1985. Methods of Structural Safety. Prentice-Hall, Englewood Cliffs, N.J.

Madsen, H.O., Torhaug, R. and Cramer, E.H. 1991. Probability-based cost benefit analysis of fatigue design, inspection and maintenance. Proceedings of the Marine Structural Inspection, Maintenance and Monitoring Symposium, SSC/SNAME, Arlington, V.A., II.E.1–12.

Maes, M.A., Fritzson, K.E. and Glowienka, S. 2006. Structural robustness in the light of risk and consequence analysis. Structural Engineering International, IABSE, 16(2): 101–107.

Marler, R.T. and Arora, J.S. 2010. The weighted sum method for multi-objective optimization: new insights. Structural and Multidisciplinary Optimization, 41(6): 853–862.

Marsh, P.S. and Frangopol, D.M. 2008. Reinforced concrete bridge deck reliability model incorporating temporal and spatial variations of probabilistic corrosion rate sensor data. Reliability Engineering & System Safety, Elsevier, 93(3): 364–409.

MathWorks. 2016. Optimization Toolbox™ User's Guide, MathWorks, Natick, MA.

Messervey, T.B., Frangopol, D.M. and Casciati, S. 2011. Application of the statistics of extremes to the reliability assessment and performance prediction of monitored highway bridges. Structure and Infrastructure Engineering, Taylor & Francis, 7(1-2): 87–99.

Miyamoto, A., Kawamura, K. and Nakamura, H. 2000. Bridge management system and maintenance optimization for existing bridges. Computer-Aided Civil and Infrastructure Engineering, John Wiley & Sons, 15(1): 45–55.

Moan, T. 2005. Reliability-based management of inspection, maintenance and repair of offshore structures. Structure and Infrastructure Engineering, Taylor & Francis, 1(1): 33–62.

Modarres, M., Kaminskiy, M.P. and Krivtsov, V. 2017. Reliability Engineering & Risk Analysis. 3rd Edition. CRC Press, U.S.
Mohanty, J.R., Verma, B.B. and Ray, P.K. 2009. Prediction of fatigue crack growth and residual life using an exponential model: Part I (constant amplitude loading). International Journal of Fatigue, Elsevier, 31(3): 418–424.
Mondoro, A., Frangopol, D.M. and Soliman, M. 2017. Optimal risk-based management of coastal bridges vulnerable to hurricanes. Journal of Infrastructure Systems, ASCE, 23(3): 4016046.
Mondoro, A. and Frangopol, D.M. 2018. Risk-based cost-benefit analysis for the retrofit of bridges exposed to extreme hydrologic events considering multiple failure modes. Engineering Structures, Elsevier, 159: 310–319.
NCHRP. 2003. Bridge life-cycle cost analysis. NCHRP-report 483, Transportation Research Board, Washington D.C.
NCHRP. 2005. Concrete bridge deck performance. NCHRP-synthesis 333, Transportation Research Board, National Cooperative Highway Research Program, Washington D.C.
NCHRP. 2006. Manual on service life of corrosion-damaged reinforced concrete bridge superstructure elements. NCHRP-report 558, Transportation Research Board, National Cooperative Highway Research Program, Washington D.C.
NCHRP. 2012. Estimating life expectancies of highway assets—Volume 1: Guidebook. NCHRP-report 713, Transportation Research Board, Washington D.C.
Neves, L.C., Frangopol, D.M. and Petcherdchoo, A. 2006. Probabilistic lifetime-oriented multiobjective optimization of bridge maintenance: Combination of maintenance types. Journal of Structural Engineering, ASCE, 132(11): 1821–1834.
Nijkamp, P. and van Delft, A. 1977. Multi-Criteria Analysis and Regional Decision-Making. Springer Science & Business Media, Leiden, The Netherlands.
Okasha, N.M. and Frangopol, D.M. 2009. Lifetime-oriented multi-objective optimization of structural maintenance considering system reliability, redundancy and life-cycle cost using GA. Structural Safety, Elsevier, 31(6): 460–474.
Okasha, N.M. and Frangopol, D.M. 2010a. Novel approach for multicriteria optimization of life-cycle preventive and essential maintenance of deteriorating structures. Journal of Structural Engineering, ASCE, 136(8): 1009–1022.
Okasha, N.M. and Frangopol, D.M. 2010b. Redundancy of structural systems with and without maintenance: An approach based on lifetime functions. Reliability Engineering & System Safety, Elsevier, 95(5): 520–533.
Okasha, N.M. and Frangopol, D.M. 2010c. Time-variant redundancy of structural systems. Structure and Infrastructure Engineering, Taylor & Francis, 6(1-2): 279–301.
Okasha, N.M., Frangopol, D.M. and Decò, A. 2010. Integration of structural health monitoring in life-cycle performance assessment of ship structures under uncertainty. Marine Structures, Elsevier, 23(3): 303–321.
Okasha, N.M. and Frangopol, D.M. 2011. Computational platform for the integrated life-cycle management of highway bridges. Engineering Structures, Elsevier, 33(7): 2145–2153.

Okasha, N.M., Frangopol, D.M., Saydam, D. and Salvino, L.W. 2011. Reliability analysis and damage detection in high speed naval crafts based on structural health monitoring data. Structural Health Monitoring, Sage Publication, 10(4): 361–379.

Okasha, N.M. and Frangopol, D.M. 2012. Integration of structural health monitoring in a system performance based life-cycle bridge management framework. Structure and Infrastructure Engineering, Taylor & Francis, 8(11): 999–1016.

Onoufriou, T. and Frangopol, D.M. 2002. Reliability-based inspection optimization of complex structures: a brief retrospective. Computers & Structures, Elsevier, 80(12): 1133–1144.

Orcesi, A.D., Frangopol, D.M. and Kim, S. 2010. Optimization of bridge maintenance strategies based on multiple limit states and monitoring. Engineering Structures, Elsevier, 32(3): 627–640.

Orcesi, A.D. and Frangopol, D.M. 2011a. Optimization of bridge maintenance strategies based on structural health monitoring information. Structural Safety, Elsevier, 33(1): 26–41.

Orcesi, A.D. and Frangopol, D.M. 2011b. Probability-based multiple-criteria optimization of bridge maintenance using monitoring and expected error in the decision process. Structural and Multidisciplinary Optimization. Springer, 44(1): 137–148.

Packman, P.F., Pearson, H.S., Owens, J.S. and Young, G. 1969. Definition of fatigue cracks through nondestructive testing. Journal of Materials, 4(3): 666–700.

Papazian, J.M., Nardiello, J., Silberstein, R.P., Welsh, G., Grundy, D., Craven, C., Evans, L., Goldfine, N., Michaels, J.E., Michaels, T.E., Li, Y. and Laird, C. 2007. Sensors for monitoring early stage fatigue cracking. International Journal of Fatigue, Elsevier, 29(9-11): 1668–1680.

Paris, P. and Erdogan, F. 1963. A critical analysis of crack propagation laws. Journal of Basic Engineering, ASME, 85(4): 528–533.

Roy, B. 1971. Problems and methods with multiple objective functions. Mathematical Programming, Springer, 1(1): 239–266.

Saaty, T.L. 1977. A scaling method for priorities in hierarchical structures. Journal of Mathematical Psychology, 15(3): 234–281.

Saaty, T.L. 2003. Decision-making with the AHP: Why is the principal eigenvalue necessary. European Journal of Operational Research, 145(1): 85–91.

Sabatino, S., Frangopol, D.M. and Dong, Y. 2016. Life-cycle utility-informed maintenance planning based on lifetime functions: Optimum balancing of cost, failure consequences, and performance benefit. Structure and Infrastructure Engineering, Taylor & Francis, 12(7): 830–847.

Sabatino, S. and Frangopol, D.M. 2017. Decision making framework for optimal SHM planning of ship structures considering availability and utility. Ocean Engineering, Elsevier, 135: 194–206.

Sánchez-Silva, M., Frangopol, D.M., Padgett, J. and Soliman, M. 2016. Maintenance and operation of infrastructure systems: Review. Journal of Structural Engineering, ASCE, 142(9): F4016004.

Saxena, D.K., Duro, J.A., Tiwari, A., Deb, K. and Qingfu Zhang. 2013. Objective reduction in many-objective optimization: linear and nonlinear algorithms. Evolutionary Computation, IEEE, 17(1): 77–99.

Saydam, D. and Frangopol, D.M. 2011. Time-dependent performance indicators of damaged bridge superstructures. Engineering Structures, Elsevier, 33(9): 2458–2471.

Saydam, D. and Frangopol, D.M. 2013. Performance assessment of damaged ship hulls. Ocean Engineering, Elsevier, 68(8): 65–76.

Saydam, D., Frangopol, D.M. and Dong, Y. 2013. Assessment of risk using bridge element condition ratings. Journal of Infrastructure Systems, ASCE, 19(3): 252–265.

Schijve, J. 2003. Fatigue of structures and materials in the 20th century and the state of the art. International Journal of Fatigue, Elsevier, 25(8): 679–702.

Soliman, M., Frangopol, D.M. and Kim, S. 2013a. Probabilistic optimum inspection planning of steel bridges with multiple fatigue sensitive details. Engineering Structures, Elsevier, 49(8): 996–1006.

Soliman, M., Frangopol, D.M. and Kwon, K. 2013b. Fatigue assessment and service life prediction of existing steel bridges by integrating SHM into a probabilistic bi-linear S-N approach. Journal of Structural Engineering, ASCE, 139(10): 1728–1740.

Soliman, M. and Frangopol, D.M. 2014. Life-cycle management of fatigue-sensitive structures integrating inspection information. Journal of Infrastructure Systems, ASCE, 20(2): 04014001.

Soliman, M., Barone, G. and Frangopol, D.M. 2015. Fatigue reliability and service life prediction of aluminum naval ship details based on monitoring data. Structural Health Monitoring, SAGE, 14(1): 3–19.

Soliman, M., Frangopol, D.M. and Mondoro, A. 2016. A probabilistic approach for optimizing inspection, monitoring, and maintenance actions against fatigue of critical ship details. Structural Safety, Elsevier, 60: 91–101.

Staszewski, W.J. 2008. Fatigue crack detection using smart sensor technologies. Fatigue & Fracture of Engineering Materials & Structures, Wiley, 31(8): 609–610.

Stewart, M.G. 2004. Spatial variability of pitting corrosion and its influence on structural fragility and reliability of RC beams in flexure. Structural Safety, Elsevier, 26(4): 453–470.

Stillwell, W.G., Seaver, D.A. and Edwards, W. 1981. A comparison of weight approximation techniques in multiattribute utility decision making. Organizational Behavior and Human Performance, Elsevier, 28(1): 62–77.

Strauss, A., Frangopol, D.M. and Kim, S. 2008. Use of monitoring extreme data for the performance prediction of structures: Bayesian updating. Engineering Structures, Elsevier, 30(12): 3654–3666.

Taguchi, G., Elsayed, E.A. and Hsiang, T.C. 1988. Quality Engineering in Production Systems. New York, NY, McGraw-Hill.

Thoft-Christensen, P. and Murotsu, Y. 1986. Application of Structural System Reliability Theory, Springer-Verlag.

Thoft-Christensen, P. and Sørensen, J.D. 1987. Optimal strategy for inspection and repair of structural systems. Civil Engineering and Environmental Systems, Taylor & Francis, 4(2): 94–100.

Torres-Acosta, A.A. and Martinez-Madrid, M. 2003. Residual life of corroding reinforced concrete structures in marine environment. Journal of Materials in Civil Engineering, ASCE, 15(4): 344–353.

van Noortwijk, J.M. and Frangopol, D.M. 2004. Two probabilistic life-cycle maintenance models for deteriorating civil infrastructures. Probabilistic Engineering Mechanics, Elsevier, 19(4); 345–359.

Verel, S., Liefooghe, A., Jourdan, L. and Dhaenens, C. 2011. Analyzing the effect of objective correlation on the efficient set of MNK-landscapes. Proceedings of the 5th Conference on Learning and Intelligent Optimization (LION 5), January 17–21, Rome, Italy.

Voogd, H. 1983. Multicriteria Evaluation for Urban and Regional Planning. Pion, London.

Wang, Y.-M. and Luo, Y. 2010. Integration of correlations with standard deviations for determining attribute weights in multiple attribute decision making. Mathematical and Computer Modelling, Elsevier, 51(1-2): 1–12.

Yeh, C.-H. 2002. A Problem-based selection of multi-attribute decision-making methods. International Transactions in Operational Research, Blackwell Publishing, 9(2): 169–181.

Yoon, K.P. and Hwang, C.-L. 1995. Multiple Attribute Decision Making: An Introduction. SAGE Publication Inc., London.

Zanakis, S.H., Solomon, A., Wishart, N. and Dublish, S. 1998. Multi-attribute decision making: A simulation comparison of select methods. European Journal of Operational Research, Elsevier, 107(3): 507–529.

Zayed, T.M., Chang, L.-M. and Fricker, J.D. 2002. Life-cycle cost based maintenance plan for steel bridge protection systems. Journal of Performance of Constructive Facilities, ASCE, 16(2): 55–62.

Zhang, J. and Lounis, Z. 2006. Sensitivity analysis of simplified diffusion-based corrosion initiation model of concrete structures exposed to chlorides. Cement and Concrete Research, Elsevier, 36(7): 1312–1323.

Zhu, B. and Frangopol, D.M. 2012. Reliability, redundancy and risk as performance indicators of structural systems during their life-cycle. Engineering Structures, Elsevier, 41(8): 34–49.

Zhu, B. and Frangopol, D.M. 2013a. Risk-based approach for optimum maintenance of bridges under traffic and earthquake loads. Journal of Structural Engineering, ASCE, 139(3): 422–434.

Zhu, B. and Frangopol, D.M. 2013b. Reliability assessment of ship structures using Bayesian updating. Engineering Structures, Elsevier, 56: 1836–1847.

Zhu, B. and Frangopol, D.M. 2013c. Incorporation of SHM data on load effects in the reliability and redundancy assessment of ships using Bayesian updating. Structural Health Monitoring, Sage Publication,13(4): 377–392.

Zhu, B. and Frangopol, D.M. 2015. Effects of post-failure material behavior on redundancy factors for design of structural components in nondeterministic systems. Structure and Infrastructure Engineering, Taylor & Francis, 11(4): 466–485.

Index

N

O

P

Chapter 1

Fig. 1.1, p. 4

Chapter 3

Fig. 3.7, p. 68

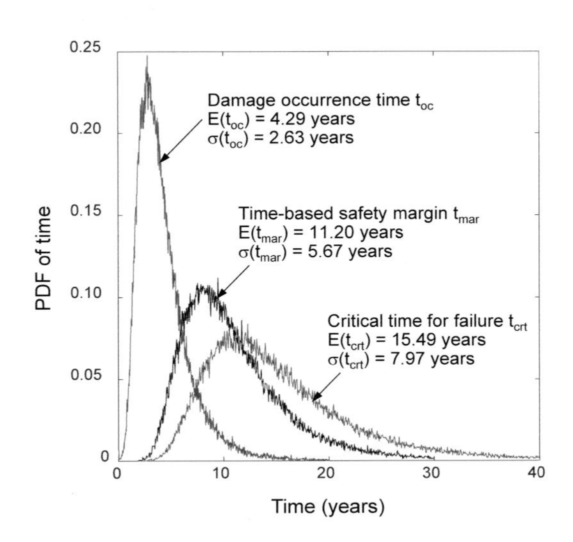

Chapter 5

Fig. 5.1, p. 99

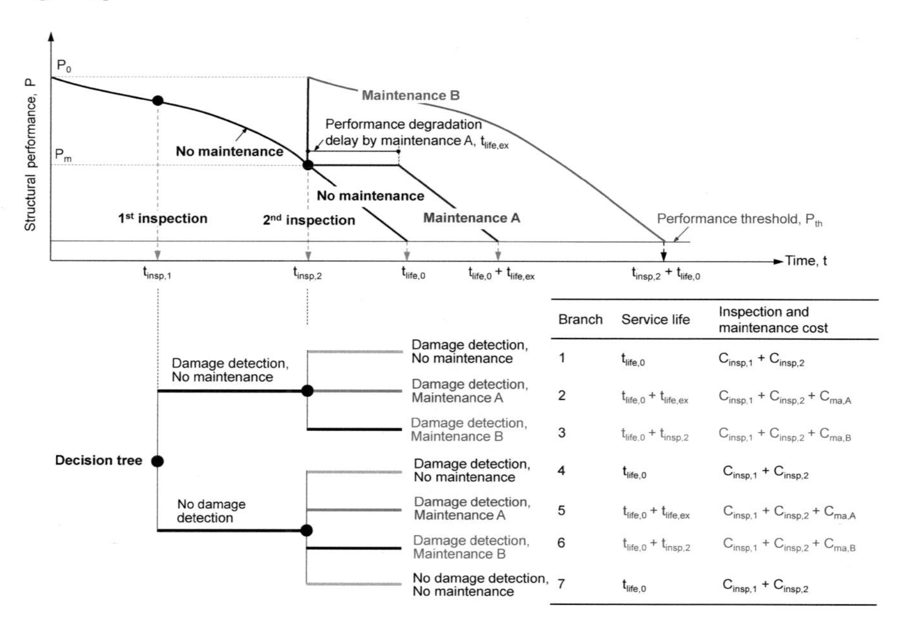

Branch	Service life	Inspection and maintenance cost
1	$t_{life,0}$	$C_{insp,1} + C_{insp,2}$
2	$t_{life,0} + t_{life,ex}$	$C_{insp,1} + C_{insp,2} + C_{ma,A}$
3	$t_{life,0} + t_{insp,2}$	$C_{insp,1} + C_{insp,2} + C_{ma,B}$
4	$t_{life,0}$	$C_{insp,1} + C_{insp,2}$
5	$t_{life,0} + t_{life,ex}$	$C_{insp,1} + C_{insp,2} + C_{ma,A}$
6	$t_{life,0} + t_{insp,2}$	$C_{insp,1} + C_{insp,2} + C_{ma,B}$
7	$t_{life,0}$	$C_{insp,1} + C_{insp,2}$

Chapter 6

Fig. 6.5, p. 128

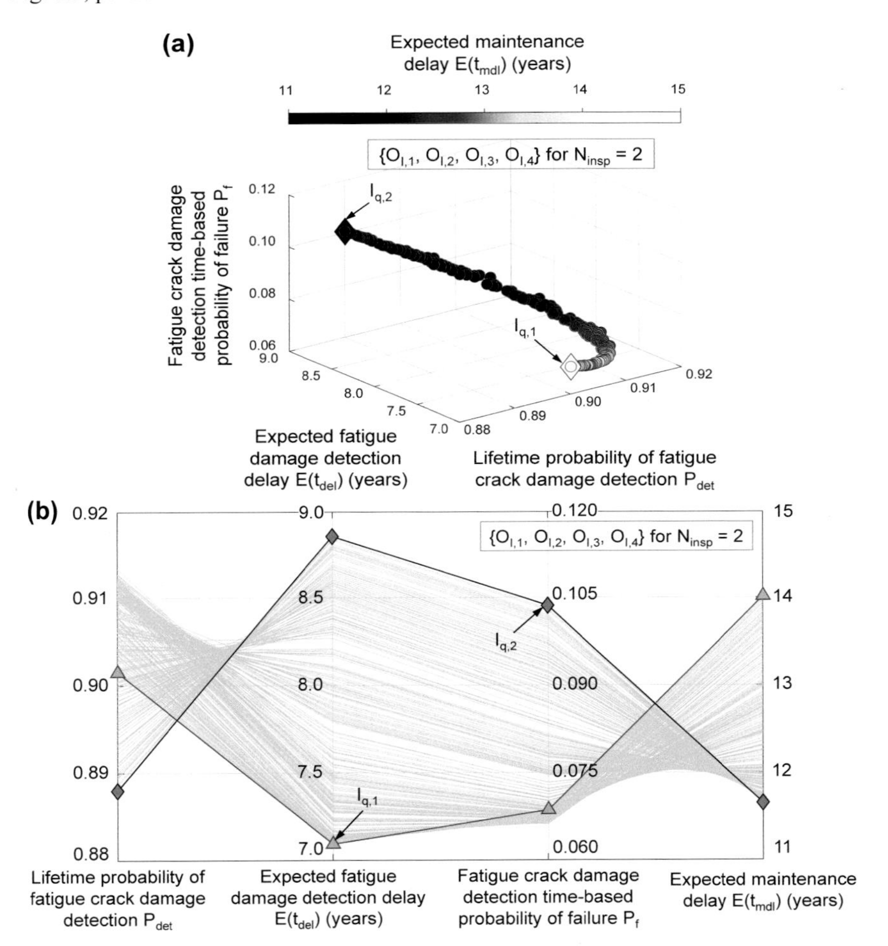

Chapter 6

Fig. 6.6, p. 129

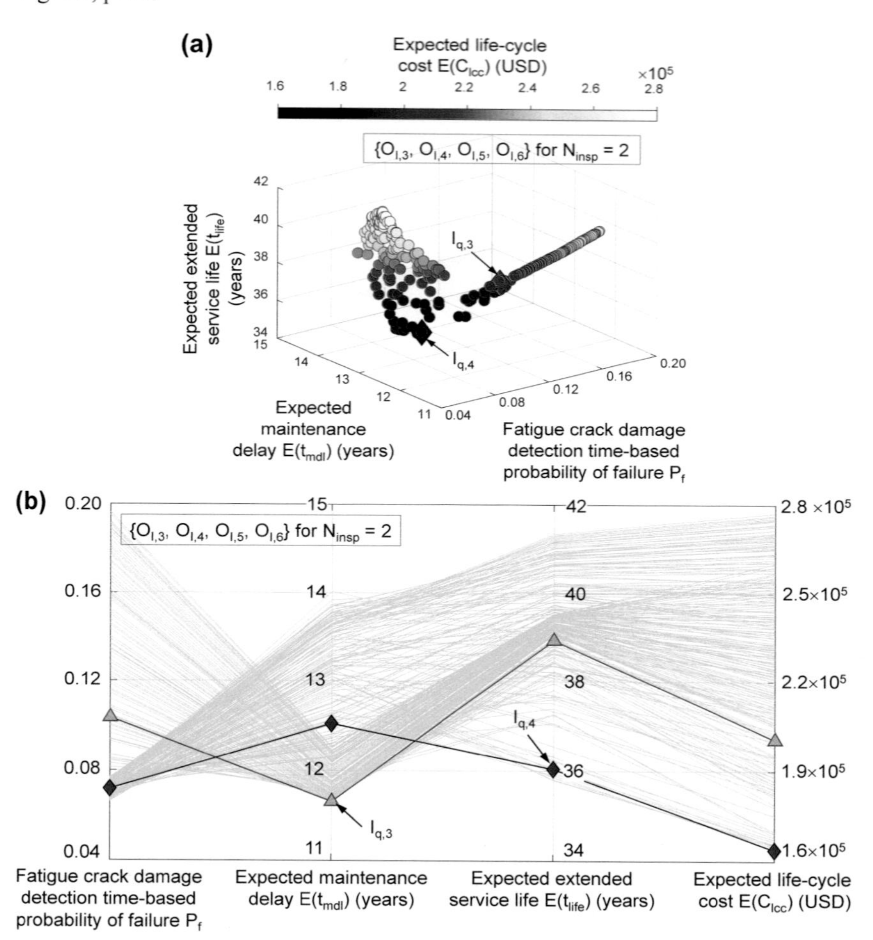

Chapter 7

Fig. 7.7, p. 168

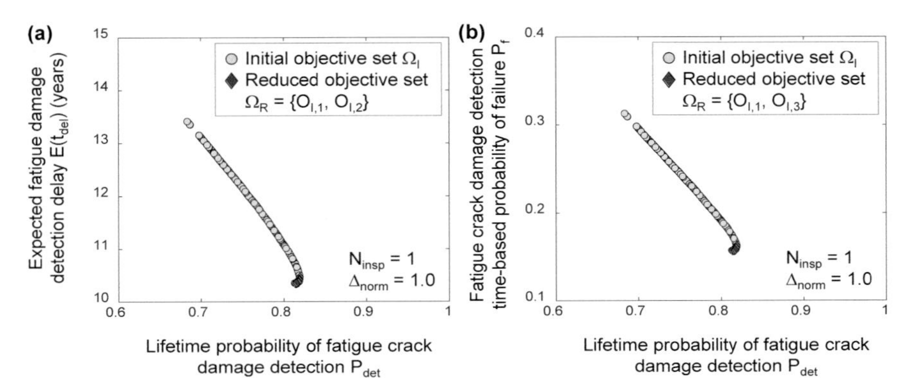

Chapter 7

Fig. 7.8, p. 168

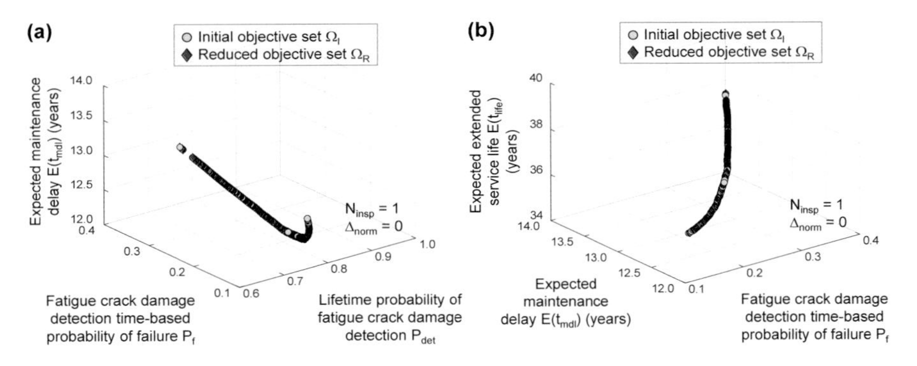

Chapter 7

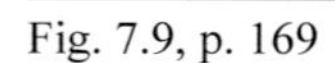
Fig. 7.9, p. 169

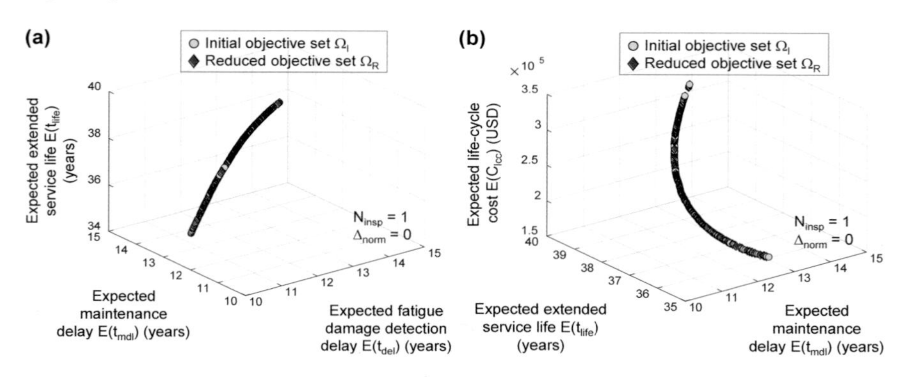

Chapter 7

Fig. 7.10, p. 171

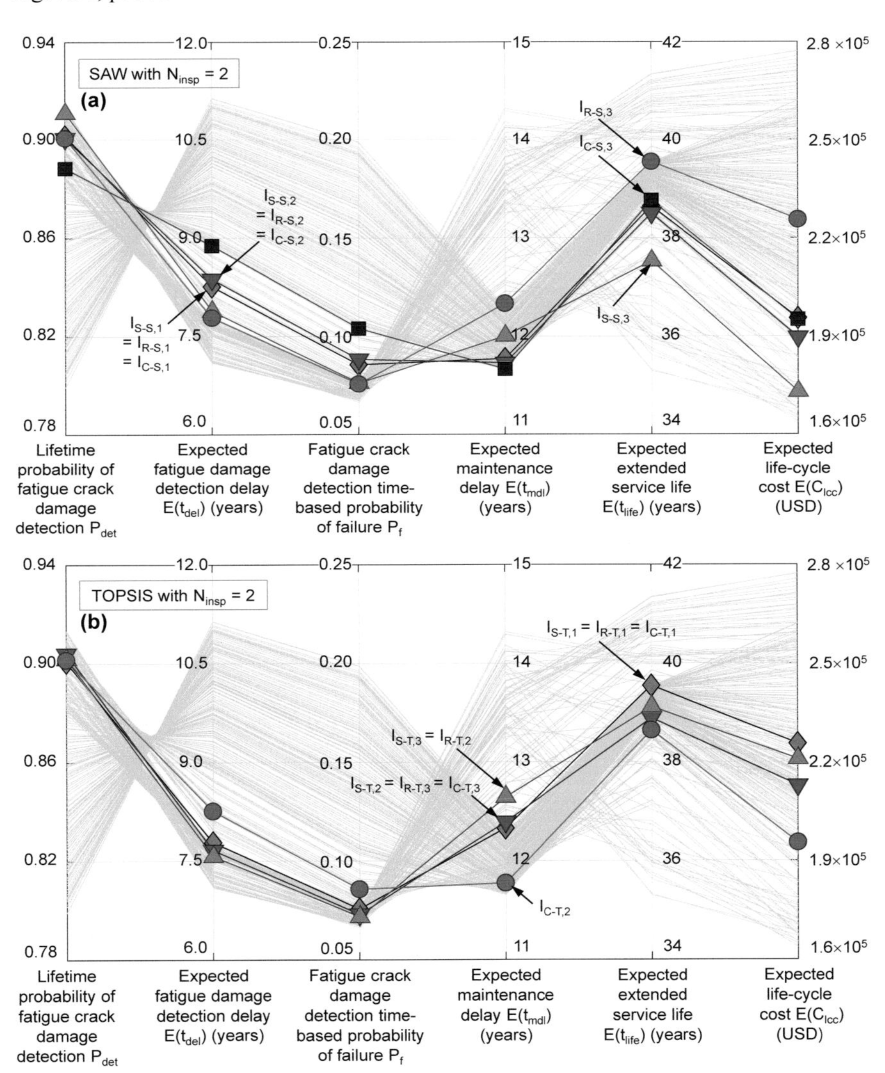

Chapter 7

Fig. 7.11, p. 173

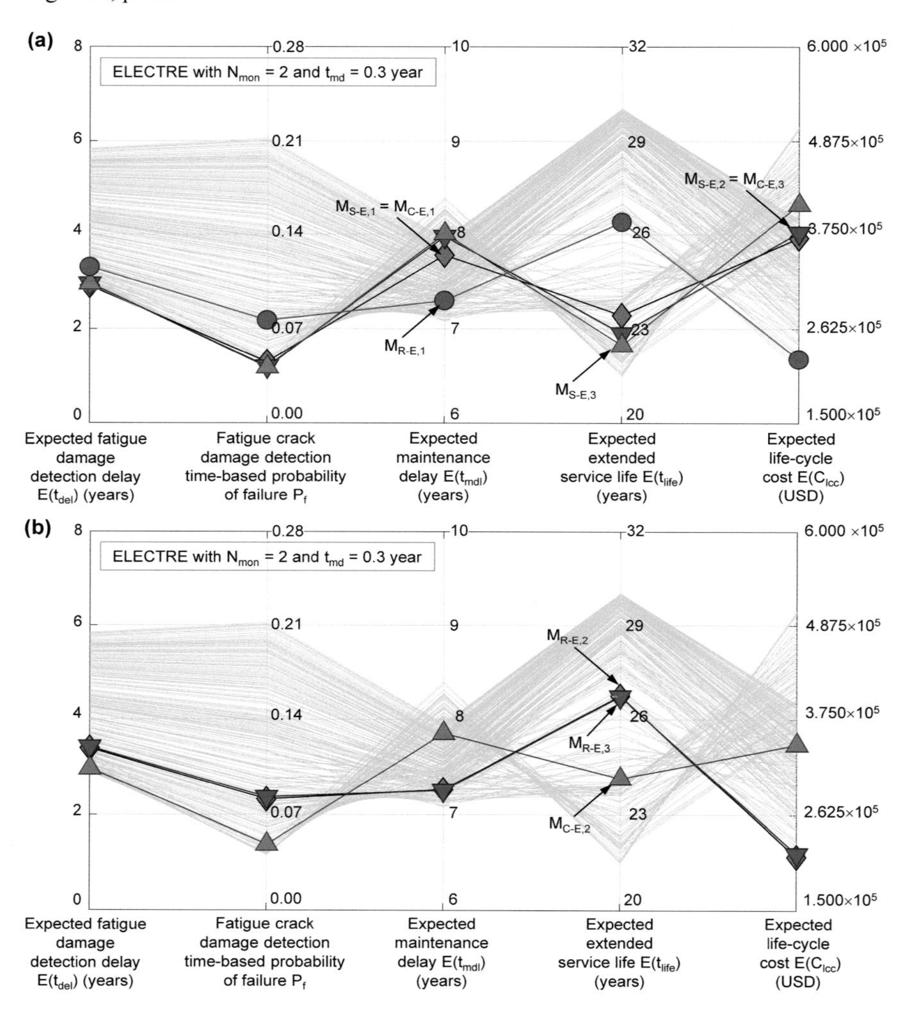